Gene Genealogies, Variation an

Gene Genealogies, Variation and Evolution

A Primer in Coalescent Theory

Jotun Hein
University of Oxford, UK

Mikkel H. Schierup

and

Carsten Wiuf
University of Aarhus, Denmark

OXFORD
UNIVERSITY PRESS

OXFORD

UNIVERSITY PRESS

Great Clarendon Street, Oxford OX2 6DP

Oxford University Press is a department of the University of Oxford.
It furthers the University's objective of excellence in research, scholarship,
and education by publishing worldwide in

Oxford New York

Auckland Bangkok Buenos Aires Cape Town Chennai
Dar es Salaam Delhi Hong Kong Istanbul Karachi Kolkata
Kuala Lumpur Madrid Melbourne Mexico City Mumbai Nairobi
São Paulo Shanghai Taipei Tokyo Toronto

Oxford is a registered trade mark of Oxford University Press
in the UK and in certain other countries

Published in the United States
by Oxford University Press Inc., New York

British Library Cataloguing in Publication Data
Data available

Library of Congress Cataloging in Publication Data
Data available
ISBN 0-19-852995-3 (hbk.)
ISBN 0-19-852996-1 (pbk.)

1 3 5 7 9 10 8 6 4 2

Typeset by Newgen Imaging Systems (P) Ltd., Chennai, India
Printed in Great Britain
on acid-free paper by
Biddles Ltd., King's Lynn, Norfolk

Preface

Coalescent theory has gone from an obscure corner of population genetics to a central concept for anybody that studies variation at the sequence level.

Besides filling the obvious need for a book on this subject, it is also our wish to present this theory in a straightforward and elementary manner that could dispel the misconception that coalescent theory is inherently very difficult and needs a strong mathematical background to understand it. The key issues needed for data analysis require only basic combinatorics and probability theory. Despite the present prominence of coalescent theory, it also belongs to the future. From an application point of view, human evolution and association mapping/fine scale mapping are two areas that are bound to grow enormously in the next few years. And to make optimal use of the coming flood of data, theoretical advances will be needed. There are areas where present theory fails (or is impractically slow) in the presence of real data and if empirical researchers are to use coalescent based method, there are plenty of challenges for the theoretician both in modelling and in improvement of simulation algorithms.

The present book is definitely not exhaustive, but is only meant to provide a good basis for further study. Chapter 1 provides the basics for understanding the assumptions behind and derivation of the basic coalescent model. Chapter 2 introduces the models of alleles and sequences and associated mutation processes, and Chapter 3 gives some examples of quantities that can be calculated on coalescent genealogies. The basic coalescent is naive for many applications and Chapter 4 and 5 relax the assumptions of the basic coalescent to allow for population size changes, population structure, various forms of selection, and recombination. Chapter 6 changes the emphasis from describing the coalescence structure to inferring parameters from data using knowledge on this structure. In Chapter 7 and 8 two areas of much current interest and where coalescent theory is likely to play a major role are introduced. Chapter 7 discusses the usage of coalescence theory in the field of linkage disequilibrium mapping that aims at locating genes underlying common diseases. Chapter 8 relates the potential and use of coalescence theory to human evolution. In an appendix a brief introduction to the web based tools developed in connection with this book are presented. The collection of tools can be found at http://www.coalescent.dk.

Acknowledgments

We thank Jeppe Warberg Larsen for design of the many illustrations of the book. Jesper Nymann Madsen is acknowledged for sharing results and material, and Thomas Christensen, Xavier Vekemans, and Gil McVean for sharing unpublished material. Thomas Bataillon, Ole Fredslund, Freddy B. Christiansen, Hans Siegismund, Bjarne Knudsen, Thomas Mailund, and Rosalind Harding provided critical and very helpful reading of parts or all of the manuscript. Lasse Westh-Nielsen and Christian Bach are thanked for helping with the generation of plots. Yun Song is thanked for implementing programs that were useful for making several illustrations and for many discussions on combinatorial aspects of coalescent theory. Victoria Hensford is thanked for correcting the spelling in many places. We also wish to thank people at the Bioinformatics Research Center at the University of Aarhus for providing a stimulating research environment, and people at the Department of Statistics, University of Oxford, for providing an ideal setting for learning about coalescent theory, in particular the groups of Peter Donnelly, Gil McVean, and Bob Griffiths. The most warm-hearted thanks go to our wives, Anne-Mette, Inger, and Marie, who have given us the time and opportunity to complete this project. The book benefitted very much from its use in teaching at the Summer Institute in statistical genetics at North Carolina State University, and graduate courses at the University of Aarhus and at the University of Oxford. We thank the Danish Cancer Society, the Aarhus University Research Foundation, and SNF in Denmark, and MRC and EPSRC in the UK for financial support.

The contributions of the three authors to this book are of different characters and are present in all chapters. The alphabetic order reflects this fact and is not a ranking of the contributions.

Contents

1 The basic coalescent

1.1 Introduction

In this chapter, we first motivate the need for a mathematical model that can describe the process generating genetic data, with specific emphasis on human variation. We assume that genetic data are in the form of DNA sequence data. The sequences or genes are all homologous copies of the same genetic region in the genome of a species. Whether one or both copies of a gene in an individual is sampled does not matter here, what matters is their number and genetic type. Such data are collected from one or several present-day populations of a single species, and from this sample we want to infer details about the evolutionary processes that created the data. (A population is here best understood as a population of genes, rather than a population of individuals, because we focus at the level of genes.) The inferential analysis is retrospective; we seek to understand aspects of the sample's (and the population's) evolutionary past through analysis of the present day sample. Below we sketch an analysis of a data set from humans to make more clear the different types of questions that coalescent theory seeks to answer.

The human genome consists of twenty-two autosomal chromosomes, the two sex chromosomes X and Y, and the mitochondrion. One representative sequence of the human genome has recently been approximately determined, and is presently subject to refinement. Additionally, a major effort has now been directed towards determining the population variation in the human genome through determination of single nucleotide polymorphism (SNPs). These are positions that vary within or between human populations. Table 1.1 shows the sizes of each chromosome and the number of positions where variation have so far been detected. The human genome and the variation that is observed is the result of interaction among evolutionary forces, such as mutational and selectional processes, mixing of variation through recombination, and demographic factors, such as the size, history, and geographical structure of the populations. Effects of demography are illustrated by the major colonisations of the globe (Figure 1.1). The present human population migrated out of East Africa approximately 100,000 years ago

Table 1.1   The human genome[a]

Chromosome	Size in Mb	Length in cM	Genes	SNPs	SNP density
1	245	293	1,945	426	1.74
2	243	277	1,283	396	1.63
3	199	233	1,049	317	1.59
4	191	212	765	318	1.66
5	181	198	879	323	1.78
6	171	201	1,053	309	1.79
7	158	184	952	282	1.78
8	146	166	717	256	1.75
9	134	167	755	263	1.96
10	135	182	756	250	1.85
11	135	156	1,294	249	1.84
12	133	169	1,006	213	1.60
13	114	118	341	166	1.46
14	105	129	647	148	1.41
15	100	110	592	166	1.66
16	90	131	900	183	2.03
17	82	129	1,121	144	1.76
18	78	124	267	138	1.77
19	64	110	1,303	110	1.72
20	64	97	631	232	3.63
21	47	60	231	88	1.87
22	49	58	485	121	2.47
X	152	198	750	45	1.61
Y	50	1	94	30	1.60

[a] The second column shows the length of each chromosome in million bases (Mb), the third the length in centiMorgans (cM), the fourth the estimated number of genes for each chromosome, the fifth the number of currently identified SNPs (thousands), and the last column shows the current density of SNPs per kilo bases (kb). The detected number of SNPs and thus also the SNP density will go up within the next few years. The data is from www.ensembl.org, release 17.33.1 (July 2003), except for the genetic map lengths which are taken from the Genethon genetic map.

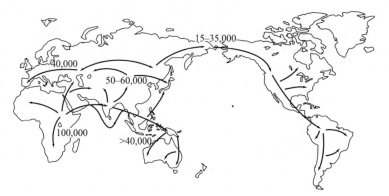

Figure 1.1   The world and historical migrations. The approximate dates of mass migrations (years ago) have mainly been determined by dating fossils found at different locations. Adapted from Cavalli-Sforza (2001).

Table 1.2 Human population growth during the last 12,000 years[a]

Time	10.000 BC	0	1750	1950	2000
Population (in millions)	6	252	771	2,521	6,055
Annual growth (%)	0.008	0.037	0.064	0.594	1.752

[a] The growth rates are point estimates at the different years.

 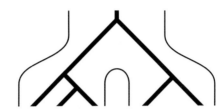

Figure 1.2 A population without geographical structure and a population consisting of two isolated subpopulations descending from a common ancestral population. The width between the thin black lines represents the size of the population at a given time in the past, such that the present time is at the bottom of the figure. The trees represent the ancestral relationship of the sampled genes. In each figure five genes are sampled, a line represents a gene's lineage as its history is back tracked. When two lines meet a common ancestor is encountered. After all genes have found a common ancestor there is just one lineage left, the lineage of the most recent common ancestor of the sample.

to colonise the world, arriving lastly at South America (for further discussion, see Chapter 8). The long term growth of the whole human population is shown in Table 1.2. It appears that the growth rate has been accelerating, that is, the human population is growing faster than exponential.

Geographical factors affect the possible patterns of genetic variation in populations. To illustrate the effect of two factors, population subdivision and population growth, we consider two simple and extreme scenarios. In Figure 1.2, the two geographical scenarios are shown—in the first there is one uniform population, where each gene has the same expected relationship to any other gene in the population (here 'gene' does not imply any knowledge about the type of the gene). The second scenario assumes that the population is made of two subpopulations of equal size that have been separated for, say, 100,000 years, but with low rate of migration between the two subpopulations. It is likely that we would find positions in the DNA-sequence that have, say, adenine (A) in most genes in subpopulation 1 and guanine (G) in most individuals in subpopulation 2. This could be the consequence of a mutation that occurred 90,000 years ago in a gene in subpopulation 2 and was randomly transmitted to most genes in this subpopulation over time. The mutant is only rarely expected to spread in subpopulation 1 unless the migration rate is high. This pattern could be repeated many times throughout the genome, reflecting the division of the two subpopulations; a pattern that would be extremely unlikely in a uniform population.

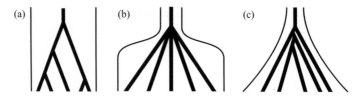

Figure 1.3   A population with (a) constant population size, (b) sudden large increase in population size (explosion), and (c) exponential growth. The width between the thin black lines represents the size of the population at a given time in the past. See the text for further explanation.

Population growth is illustrated by Figure 1.3. Three growth scenarios are shown. In the first, the population size is constant arbitrarily far back in time. In the second case, the population has quickly jumped from being very small to being very large some time ago. The conclusion we can draw about the ancestry of the sample of genes in this case depends on the relative sizes of the population before and after the explosion (time is here running in the usual direction, from the past towards the present time), and on when the explosion happened. If it happened too far back in time the genealogy of a sample will look like the genealogy in the first case, because all genes in the sample will find common ancestors after the explosion. At the other extreme it happened in recent times and all genes have distinct lineages at the time of the explosion. If the population size before the explosion is relatively small, most lineages would collapse instantly just before the explosion. Between these two extremes the genealogy will look like a distorted version of a genealogy from the population in the first case, depending on the sizes of the population before and after the explosion.

In the third case, the population is assumed to grow exponentially, that is, the population size decreases exponentially when we go back in time. This case has properties intermediate between the first two cases: There would be short internal branches just after the most recent common ancestor (MRCA) of the sample. However, we should not expect to see a sudden collapse of lineages as in the second case, because the population size is continuously decreasing. All these conclusions cannot be made conclusively without a complete specification of the model and we stress that they depend on the population sizes through time, how fast the sizes decreases as we go back in time, etc.

The genealogy influences the type of the genes observed in the sample. When a gene is passed on from parent to offspring there is a chance (however small) that the transmitted gene is a mutated copy of the parental gene. If the genealogy of a sample spans many generations there is a higher chance of seeing different types of genes than if the genealogy spans few generations. Also the shape of the genealogy is of importance: In Figure 1.3 mutations in (b) and (c) tend to produce singletons (genetic types that only occur in one copy in the sample), in contrast to (a) where it is much more likely that a mutation is found in several members of the sample.

The lesson from these simple examples is that the population scenario has consequences for the probability of the observed data set. Mathematical models will give data exact probabilities as a function of underlying parameters describing population history.

1.2   A Y-chromosome data set

If we zoom in on a small fraction of the Y-chromosome (see Figure 1.4), we may look for variation between individual chromosomes in human populations. The Y-chromosome is easy to study since males carry only one copy and there is no recombination. Michael Hammer and coworkers have determined positions of variation along the chromosome and typed these in more than 2000 male individuals around the world. Figure 1.5 shows a subset of this data set. Each column is a segregating site, that is, a position that has two or more types in the sample. For simplicity, we encoded the data as 0s and 1s instead of the actual base pairs (A, T, C, and G), because at most two base pairs are present in each position. Sometimes it is possible to tell (or estimate) which of the two states is the oldest, the ancestral, and which is the youngest, the mutant. If this is the case we use zero as the ancestral state. For this data set, the ancestral state has been determined by comparison with a chimpanzee sequence. Figure 1.5 shows

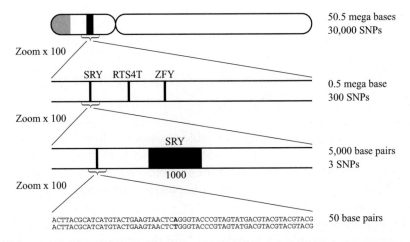

Figure 1.4   Zooming in on a small fraction of the Y-chromosome where a single nucleotide polymorphism is found (in the sex-determining gene, SRY). This polymorphism is one of the nineteen polymorphisms studied by Hammer et al. (2001) and shown in Table 1.3. The lower panel shows a place of polymorphism within a 50 bp interval.

Haplotype	1	2	3	4	5	6	7	8	9	10	11	12	13	14	15	16	17	18	19	European	North Africa	South Asia	South Africa	Total
A	0	0	0	0	0	0	0	0	0	0	0	0	0	0	0	0	0	0	0		10		47	57
B	0	1	0	0	0	0	0	0	0	0	0	0	0	0	0	0	0	0	0				41	41
C	0	1	0	0	0	0	1	0	0	1	0	1	0	0	0	1	0	0		1		4		5
E	0	1	0	0	0	0	1	0	1	0	0	0	0	0	0	0	0	0	0	33	70		153	256
F	0	1	0	0	0	0	1	1	0	0	0	0	0	0	0	0	0	0	0	6	1	19		26
G	0	1	0	0	0	0	1	0	0	1	0	1	0	0	0	0	1	0	1	24	1	5		30
H	0	1	0	0	0	0	1	0	0	1	1	0	0	0	0	0	0	0	0			27		27
I	0	1	0	0	0	0	1	0	0	1	0	1	0	0	1	0	0	0		47				47
J	0	1	0	0	0	0	1	0	0	1	0	1	1	1	1	0	0	0	0	51	35	21	1	108
K	0	1	1	1	0	0	0	0	0	0	0	0	0	0	0	0	0	0	0	4	6		1	11
L	0	1	0	0	0	0	1	0	0	1	0	1	0	0	0	1	1	0		10				10
N	1	0	0	0	0	0	0	0	0	0	0	0	0	0	0	0	0	0	0	14		10		24
P	0	1	1	0	1	1	0	0	0	0	0	0	0	0	0	0	0	0	0			3		3
Q	0	1	0	0	0	0	1	0	0	0	0	0	0	0	0	0	0	0	0	5		3		8
R	0	1	0	0	0	0	1	0	0	1	0	0	0	0	0	0	0	0	0	160	8	41		209
Consensus	0	1	0	0	0	0	1	0	0	1	0	0	0	0	0	0	0	0	0	355	131	133	243	862

Figure 1.5 Y-chromosome variation in four human population. There are nineteen segregating sites arranged in fifteen haplotypes named according to Hammer et al. (2001). The last four columns show the frequencies of the haplotypes in Europe, Northern Africa, sub-Saharan Africa, and South Asia. Missing entries indicate that a haplotype is not present. Zero denotes the estimated ancestral state of a segregating site as determined by comparison with a chimpanzee sequence. The bottom row is the consensus sequence, the sequence composed of the most frequent base in each column.

Table 1.3 Summary statistics for the Y-chromosome data set of Hammer et al. (2001)[a]

Population	Sample size	Segregating sites	Pairwise difference
European	355	16	2.48
North Africa	131	13	2.39
South Asia	133	14	2.56
South Africa	243	10	1.65

[a] 'Pairwise difference' is the average number of differences between two Y-chromosomes, for example, 2.48 is the average over $\binom{355}{2} = 62{,}835$ pairs of Y-chromosomes.

patterns of variation at nineteen variable positions in four human populations. The variation is arranged into fifteen distinct haplotypes. A haplotype is a sequence variant of the piece of the chromosomes being investigated. It can immediately be seen that different haplotypes predominate in different populations.

Table 1.3 shows summary statistics for the data set, including the number of segregating sites in each population, and the average number of differences between two chromosomes in each population. The results indicate that the South African sample has less variation than the other samples. The average number of differences between two chromosomes taken from different populations is 3.18, which is larger than the within population differences. This suggests that there is genetic differentiation between the populations.

We used the computer program Genetree by Bahlo and Griffiths (2000) to estimate different quantities from the data set. This is possible when the data set satisfy the assumptions of the infinite sites model (Chapter 2) and no recombination (Chapter 5), which is the case for the present data set. The infinite sites assumption states that at most one mutation has happened in each position. The fact that only two types, 0 and 1, are present in each position, tell us that there are no obvious contradictions to this assumption. Combining the infinite sites assumption with the assumption of no recombination implies that the haplotypes can be explained or depicted by a gene tree, a graphical representation of the data that connects any two haplotypes in the sample with a series of mutation events, one for each variable site found in the two haplotypes. In this case the gene tree can be rooted because the state of the MRCA has been estimated from a chimpanzee sequence. Due to the convention that 0 represents the ancestral state, the root sequence is the haplotype with zeros only. The root sequence differs from the consensus sequence in three places.

Figure 1.6 shows the gene tree with mutations marked by their numbers (in Figure 1.5). The gene tree is uniquely determined by the data. When the root is given, we can read off events from the gene tree. For example, if we follow the lineage from the root down to haplotype F we encounter a split into three lineages, then a mutation event (mutation 2), a split into three lineages immediately after the mutation event, a mutation event (7), a split event, and finally a mutation event (8). In comparison of lineages the ordering of events along distinct lineages cannot be determined, only the relative order of events along a single lineage is known.

Using coalescent methods it is possible to estimate the time of events, depending on the assumptions of the underlying mathematical model. In this example we used a model with four distinct subpopulations, each exchanging migrants with a scaled rate 1.0 (see Chapter 4 for details). Further, each subpopulation has equal constant population size. The scaled rate is twice the number of migrants arriving in a subpopulation per generation. In the gene tree in Figure 1.6, we see that mutations found in more than one subpopulation or found in more than one gene are estimated to be older than those found in one subpopulation or in one copy only.

The probability of the observed data can be calculated as a function of parameters in the model, such as the scaled mutation rate, or scaled migration rates. We will see how this can be done in Chapter 2. Calculating the probability over the range of parameter values, one can draw the likelihood curve. The parameter value that corresponds to the maximum of the likelihood curve is the maximum likelihood estimate of the parameter, that

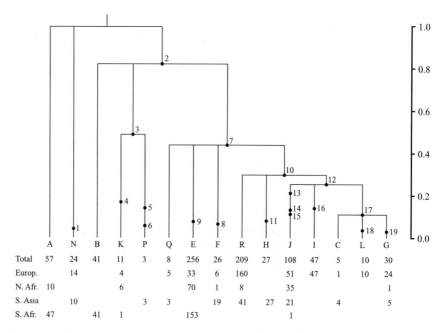

	A	N	B	K	P	Q	E	F	R	H	J	I	C	L	G
Total	57	24	41	11	3	8	256	26	209	27	108	47	5	10	30
Europ.		14		4		5	33	6	160		51	47	1	10	24
N. Afr.	10			6			70	1	8		35				1
S. Asia		10			3	3			19	41	27	21		4		5
S. Afr.	47		41	1			153					1			

Figure 1.6   Gene tree estimated from the data set in Figure 1.5 using the Genetree program. The vertical axis shows time relative to the estimated time of the root.

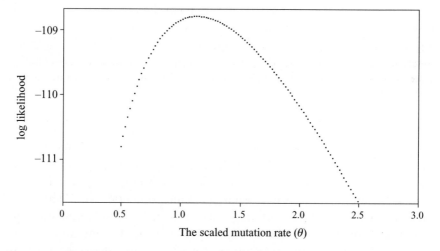

Figure 1.7   The likelihood curve for the scaled mutation rate $\theta = 4Nu$, where u is the mutation rate per generation and N the population size. The maximum likelihood estimate is approximately 1.2. The y-axis is on $\log_{10}$ scale.

is, the value of the parameter that makes the observed data under the chosen model most probable.

Figure 1.7 shows the likelihood surface of the scaled mutation rate θ, which is defined as $4Nu$, where N is the population size and u is the mutation rate per generation (see Chapter 2). This curve was calculated under the same model as discussed above with fixed migration rates and fixed number of subpopulations. The maximum likelihood estimate of θ is approximately 1.2.

Another quantity of interest is the time of the MRCA of the sample (termed the TMRCA), that is, when was the first time the sample had one ancestor only. The probability density of this quantity given the data is termed the posterior density of the TMRCA. It is shown in Figure 1.8; again the same assumptions are used. Obviously, the haplotype of the

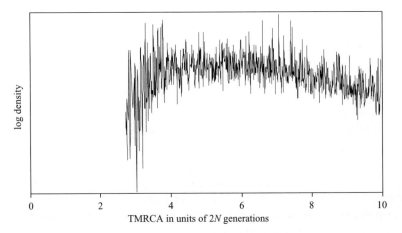

Figure 1.8   Posterior density for the TMRCA for the Y-chromosome data set. The maximum posterior estimate (the value of TMRCA that corresponds to the maximum of the curve) of the TMRCA is approximately six (after the curve has been smoothed). The posterior density is very flat which indicates that the maximum posterior estimate has a wide error margin.

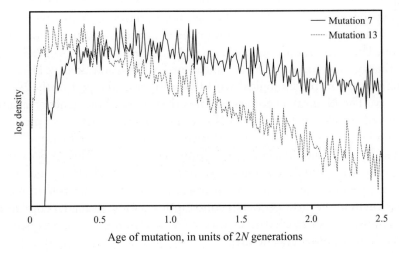

Figure 1.9   Posterior densities for the ages of mutation 7 and mutation 13 of the Y-chromosome data set.

MRCA consists of only zeros, so the TMRCA is at least as old as the most recent occurrence of the haplotype of the MRCA. Figure 1.8 shows that the TMRCA cannot be determined with great precision from this data set under the given model.

While Figure 1.6 showed the estimated relative ages for each mutation, Figure 1.9 shows the posterior density for the ages of two specific mutations. Clearly, the distributions are distinct with mutation 7 estimated to be the oldest in accordance with the point estimate embedded in Figure 1.6.

1.3   Data and theory

The main message from the Y-chromosome example is that explicit probability models of mutations, geography, and reproduction structures allow us to make statements about likely parameter values, ancestral events, dating of events, etc. Without such models, none of this is possible. The above examples are only illustrations of some simple questions one could ask. The relevant questions are both qualitative and quantitative in nature. Examples of qualitative questions are: Does the data show sign of population structure, recombination, population growth, or selection? Quantitative questions aim at estimating parameters, such as the scaled rate of recombination or migration, or the age of a mutation.

The theoretical field underlying the analysis of population data is *population genetics*. This field and many of its major theoretical results are quite old. It was pioneered by three major founding fathers: Sewall Wright (1889–1988), Ronald A. Fisher (1890–1962) and J. B. S. Haldane (1892–1964) during the 1920s and 1930s. They set the outlines for a prospective theory of the fate of genetic variation under migration, selection and random genetic drift. Later, the contributions of Motoo Kimura (1924–1994) made the theory more rigorous by advanced use of diffusion theory. He caused a major shift in the biological world view by introduction of the Neutral Theory. The Neutral Theory postulates that most of the genetic variation observed is selectively neutral. This does not imply that new mutations cannot be selected against, but that such variation is normally rapidly eliminated from the population by selection. The Neutral Theory does imply that only a very small fraction of new mutations are selectively advantageous. It provoked debate because it offered a much smaller role to natural selection than the most prevalent contemporary view; instead it emphasised the importance of stochastic factors such as variation in the frequency of an allele by random genetic drift.

The perspective in the field shifted further in the 1970s and 1980s when the emphasis changed from prospective (looking forward in time) to retrospective (looking backward in time) methods of analysis. This development was a natural consequence of the availability of genetic data sampled at the present time but shaped by past processes. These processes were of interest to make inferences about.

Theoretical population genetics has since the late 1990s been graced by increased attention from functional biology due to the appearance of completely sequenced genomes and associated data on population variation at the sequence level, turning population genetics into population genomics. This recent rise in importance stems both from the potential of mapping characters to genes (association mapping) and the possibility of a pharmacology tailored to the individual, when knowing the genotype of an individual is crucial for predicting drug metabolism and drug response. It is still an open question if these new fields can live up to the expectations, but almost irrespective of the final outcome, theoretical population genetics is bound to be central in functional biology to a degree that could not have been foreseen a decade ago.

The central approach of genealogical data analysis is a stochastic characterisation of the genealogies that relate the sequences. Evaluating the probability of a given data set then consists of two steps: First, model reproduction in the population which leads to a probabilistic description of the genealogical relationship of the sampled data. Second, each genealogy will generate the data with a specific probability when combined with a model of the mutation process.

1.4 The Wright–Fisher model

A simple model of populations describing the genealogical relationship among genes is that introduced by Wright (1931) and Fisher (1930). This basic model of reproduction provides a dynamic description of the evolution of an idealised population and the transmission of genes from one generation to the next. By genes we refer to a material entity transmitted from one generation to the next. If two copies of a gene can be distinguished we refer to them as different alleles. A sequence is a gene where the nucleotides have been determined. Thus, two sequences of the same gene are different alleles if they are not identical. The basic properties of the model in haploid and diploid versions are illustrated in Figures 1.10 and 1.11.

To facilitate comparison of haploid and diploid models we may assume a population size of $2N$ genes, corresponding to N diploid or $2N$ haploid individuals. Thus, haploid reproduction is modelled assuming $2N$ individuals. Note that other treatments of the Wright–Fisher model may assume N genes instead of $2N$ and that results therefore may differ by a factor of two reflecting this. In the haploid model, each of the genes of generation $t + 1$ are

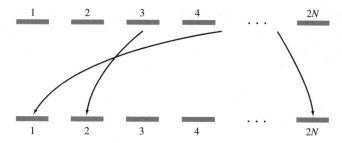

Figure 1.10   Haploid reproduction model. The genes making up the present generation (lower line) are drawn randomly with replacement from the parental generation.

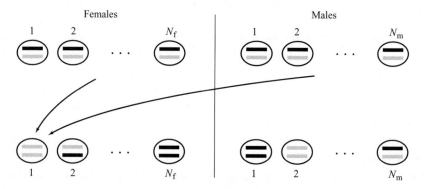

Figure 1.11   Diploid reproduction model. An individual in the present generation (lower line) draws randomly with replacement one of its genes from the female population and the other gene from the male population.

found by copying the gene of a random individual from generation t. This is repeated independently until $2N$ genes have been sampled (Figure 1.10). Each gene in generation $t + 1$ will thus have one parent gene in generation t, but it is a random one. A gene in generation t might not have any descendants in generation $t + 1$ and consequently its lineage has died out.

Diploid reproduction in the case of species with separate sexes assumes two subpopulations—females and males—of sizes N_f and N_m, respectively, with $N = N_f + N_m$, representing again $2N$ genes. Individuals in generation $t + 1$ are created one by one. Each individual chooses a male (father) and a female (mother) from the male and female populations, respectively, from generation t. Within the father and the mother one of the two genes is chosen with probability 0.5 each. This reproductive scheme is illustrated in Figure 1.11. Like the haploid model, each gene has one parent gene (in a male or a female), but each individual has two parents.

If the ancestry of two genes is traced back through time in the haploid model and the diploid model, respectively, there are certain restrictions on which parents a gene could choose in the diploid model relative to the haploid model. In the haploid model all genes choose independently of each other, while in the diploid model the second gene must choose a different parent from the first gene. Genealogies in the diploid model and the haploid model are probabilistically similar for large choices of N, N_f, and N_m, if adjustments are made to how time is scaled. This will be taken up in Section 1.10. When nothing else is stated we assume the haploid model for convenience.

1.4.1 Assumptions of the Wright–Fisher model

A number of idealised and simplifying assumptions are explicitly and implicitly made in the Wright–Fisher model of reproduction. These are:

1. *Discrete and non-overlapping generations.* In the case of humans, this is equivalent to assuming that everybody has the same lifetime expectancy from conception to reproduction (about 25 years), and that reproduction and death is simultaneous for all individuals and synchronous among all individuals. Fortunately, it turns out that the assumption is of little practical consequence. Models assuming overlapping generations (not all genes give birth or die at the same time) give probabilistically similar genealogies.

2. *Haploid individuals or two subpopulations (males and females).* In problems that do not relate to selection involving heterosis, it usually has little quantitative consequence to assume a haploid population of size $2N$ in place of a diploid population with N individuals, as noted previously.

3. *The population size is constant.* This is a genuine biological assumption and important quantities of the model will be different if the population is growing, shrinking, oscillating or has gone through a transient very small size, termed a population bottleneck (see Chapter 4).

4. *All individuals are equally fit.* This is a convenient assumption in the introduction of basic concepts, but presumably not realistic for all loci. Relaxing this assumption will be discussed in Chapter 4, most notably, because major questions in genealogical data analysis seek to investigate the presence and strength of natural selection.

5. *The population has no geographical or social structure.* Choosing parents randomly as in the Wright–Fisher model is not a realistic mechanism of reproduction in any real population. Population structure of any kind may greatly affect genealogies (Chapter 4) and this assumption is therefore important for analysis of many real data sets.

6. *The genes (or sequences) in the population are not recombining.* This is an important assumption that needs to be relaxed when analysing many real data sets. Recombination potentially occurs in most sequences, with Y-chromosome, and perhaps mitochondrial DNA, as the primary

exceptions. Unfortunately, relaxing the assumption makes analysis much more mathematically complex, mainly because the sequence sample is no longer related by a genealogical tree but rather a graph (the ancestral recombination graph, see Chapter 5) or a collection of trees.

1.4.2   The number of descendants of a gene in one generation

The number of descendants of a particular gene, i, in generation t is a stochastic variable. Its distribution is straightforward to calculate, since each time a new gene in generation $t + 1$ is created it has probability $1/(2N)$ of picking the parent i in generation t and this sampling is performed repeatedly $2N$ times with replacement.

Let v_i be the number of descendants of gene i in generation t, $i = 1, 2, \ldots, 2N$, then

$$P(v_i = k) = \binom{2N}{k} \left(\frac{1}{2N}\right)^k \left(1 - \frac{1}{2N}\right)^{2N-k}. \tag{1.1}$$

This is an example of the binomial distribution, $\mathrm{Bi}(m, p)$, with parameters $m = 2N$ and $p = 1/(2N)$. Thus, the number of genes descending from a given gene is binomially distributed. The moments of the binomial distribution are well-known: For v_i the mean is

$$E(v_i) = mp = 2N\frac{1}{2N} = 1, \tag{1.2}$$

and the variance is

$$\mathrm{Var}(v_i) = mp(1 - p) = 2N\frac{1}{2N}\left(1 - \frac{1}{2N}\right) = 1 - \frac{1}{2N}. \tag{1.3}$$

That the mean is one is a consequence of the population size being constant: If the mean number of descendants of a gene were larger/smaller than one the population would be increasing/decreasing in size. The covariance of the offspring numbers for two genes i and j is

$$\mathrm{Cov}(v_i, v_j) = E(v_i v_j) - E(v_i)E(v_j) = -\frac{1}{2N}, \tag{1.4}$$

and the correlation coefficient is

$$\mathrm{Cor}(v_i, v_j) = \frac{\mathrm{Cov}(v_i, v_j)}{\sqrt{\mathrm{Var}(v_i)\mathrm{Var}(v_j)}} = -\frac{1}{2N - 1}. \tag{1.5}$$

Thus, v_i and v_j are almost independent of each other for large $2N$. Intuitively, a negative covariance (or correlation) is expected because if gene i leaves many descendants in the next generation then j is more likely to leave few. This is because the total number of descendants of all genes in a generation is $2N$. Naturally, this effect is more pronounced in small populations than in large populations.

If $2N$ is large then v_i is almost Poisson distributed, Po(1),

$$P(v_i = k) \approx \frac{1}{k!}e^{-1} \tag{1.6}$$

with mean one and variance one. The probability that a gene does not leave descendants is $P(v_i = 0) = e^{-1} \approx 0.37$ and approximately a fraction of $1-e^{-1} \approx 0.63$ of all genes have descendants. Thus, in a large randomly mating population the present day population descends from a relatively small fraction of genes a few generations ago, namely approximately 0.63^t if t generations ago. For example, a population of size $2N = 10,000$ originates from about ten ancestral genes (0.1% of the total population) approximately fifteen generations ago ($10,000 \cdot 0.63^{15} \approx 10$). The lineages of the remaining genes (approximately $10,000 - 10 = 9990$) in the ancestral population fifteen generations ago did not survive until the present day generation.

As soon as the number of ancestral genes becomes small these calculations are no longer valid because they are based on large sample size properties. This is where the coalescent process comes in.

1.4.3 An example

Figure 1.12 shows the Wright–Fisher model of reproduction in a population of size 10 for fifteen reproduction cycles, corresponding to sixteen generations. Each gene is arbitrarily labelled with the position they have in the row (generation) they occupy. Each gene is linked to its ancestor gene in the previous generation. This diagram completely describes the genealogical relationships of all genes during these sixteen generations. Since the labels of the individuals are arbitrary, they can be sorted in each generation, such that all the children of the first parent comes first, then the children of the second parent, etc. This has been done in Figure 1.13. The embedded tree structure now becomes much more apparent, since it is easy to track the ancestry of any set of genes chosen anywhere in the diagram. It is a consequence of the model that any set of genes in a finite number of generations will have only one ancestor. It is quite possible that more than sixteen generations are needed to find the first common ancestor of all genes (the MRCA), even though this is not the case in this example. The concept of a MRCA of the whole population of genes occurring at some time is a 'back in time' statement that has a 'forward in time' consequence: At a certain generation all the lineages starting from the $2N$ genes will die out except for one. To see this, sample the whole population and trace their ancestry back

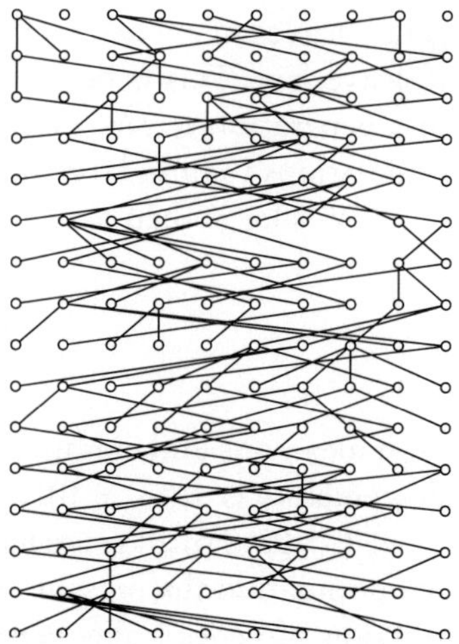

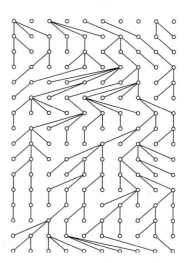

Figure 1.12   The haploid Wright–Fisher model with ten genes applied for sixteen generations, corresponding to applying the haploid model fifteen times starting with the population at the top row.

Figure 1.13   The same diagram as in Figure 1.12 but with lineages sorted such that the tree-like nature of the reproductive structure emerges.

until their MRCA has been found. All the other genes in that generation do not have any descendants in the most recent generation.

From the point of view of data analysis, only a sample of n (typically n is much smaller than $2N$, i.e. $n << 2N$) genes are taken from the present population and the genealogical ancestry of this sample is of interest. In Figure 1.14 three genes (1, 2, and 3, which are the first, third, and ninth of the whole population in Figure 1.12) have been sampled randomly in the present population. Edges back in time tracking the ancestors of these three sequences are highlighted. Two generations back in time, genes 1 and 2, find a common ancestor and this lineage might be labelled (1, 2) to reflect this fact. Seven generations further back in time back (1, 2) finds a common ancestor with 3 and the genealogical relationships of the three genes are now fully described. It is possible to follow the MRCA of all three genes further

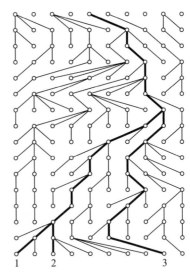

Figure 1.14   The genealogy of three randomly sampled sequences named 1, 2, 3 from left to right. The ancestry of the sequences are marked by bold lines sixteen generations back in time; nine generations back all three sampled sequences have found a common ancestor.

back in time, but there will never be any information on what happened to this MRCA further back in time from the sampled genes. Questions about the structure of genealogies of a population or a sample from a population are our main interest to address, for example:

1. What is the length of the epoch when there were only k lineages?
2. What it is the probability that the number of lineages shrinks by more than one in one generation?
3. What is the general structure of the genealogies generated by this process?

Getting answers to these question will facilitate the interpretation of actual sequence data.

1.5   The geometric distribution

Ubiquitous in population genetics and in coalescent theory in particular is the *Markov property*, which states that the probability of a specific next step in a discrete or continuous process only depends on the present state of the process, that is, the process is without memory of events prior to the present. In genetics it is natural to assume that the probability that something is going to happen (e.g. a mutation or finding a common ancestor) depends only on the situation at present (e.g. the present nucleotide, nucleotide sequence, or number of genes). In processes where time is discretely measured (e.g. in generations) the Markov property is closely associated with the *geometric* distribution. If time is continuous (e.g. as measured by a stopwatch) the analogous distribution is the *exponential* distribution. The exponential distribution will be discussed in the next section.

In analogy with the binomial distribution, the geometric distribution can be obtained by independent and repeated experiments with two possible outcomes, for example, tossing a coin, or observing whether or not a dice shows six. The binomial distribution describes the outcome of repeating the experiment n times, where each experiment has a probability p of success, and then asking for the total number of successes. The geometric distribution also describes results from a repeated experiment. However, rather than describing the number of successes, it records the time (i.e. the number of trials) until the first success.

Thus, let $X_i, i = 1, 2, \ldots$, be a series of independent and identically distributed experiments with probability p of success and $1 - p$ of failure. In population genetics this could be going back in time and in each generation observing whether two genes had found a common ancestor. Denote success with one and failure with zero. Let T be the waiting time until the first success, that is, $T = \min\{i \mid X_i = 1, i = 1, 2, \ldots\}$. This leads to the geometric probability distribution,

$$P(T = j) = (1 - p)^{j-1} p \tag{1.7}$$

$j = 1, 2, \ldots$, since $T = j$ implies $j - 1$ failures followed by one success. We take $T \sim \text{Geo}(p)$ to mean that T is geometrically distributed with parameter p.

A series of simple properties of geometrically distributed variables can easily be derived. Assume that $t_2 > t_1$. Then

$$P(T > t_2 \mid T > t_1) = P(T > t_2 - t_1), \tag{1.8}$$

$$E(T) = \frac{1}{p}, \tag{1.9}$$

and

$$\text{Var}(T) = \frac{1 - p}{p^2}. \tag{1.10}$$

Let S be a second geometrically distributed variable, $S \sim \text{Geo}(p')$. Assume S is independent of T, then

$$\min(S, T) \sim \text{Geo}(p + p' - pp'). \tag{1.11}$$

Property (1.8) illustrates the lack of memory: Knowing that T has not occurred at time t_1 does not change the probability of when it will occur

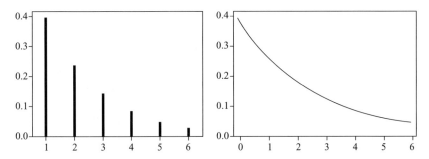

Figure 1.15   Comparison of the continuous exponential density and the discrete geometric density with mean 2.5.

in the future. An example of a geometric distribution with mean 2.5 is displayed in Figure 1.15. Note the long tail of the distribution, and the general fact that the probabilities are decreasing for increasing j.

1.6   The exponential distribution

The exponential distribution can be defined in a variety of ways. It arises naturally from the Markov property, but it can also be obtained as the limit distribution of a series of geometric distributions measured on a finer and finer ladder of time points. One then waits for an event (success) on a continuous line.

Let U have an exponential distribution, that is,

$$P(U \le t) = 1 - e^{-at}. \tag{1.12}$$

The density function of the distribution is found by differentiation of $P(U \le t)$ with respect to t,

$$f(t) = \frac{dP(U \le t)}{dt} = ae^{-at}. \tag{1.13}$$

The exponential distribution is characterised by one parameter only, the intensity a which can be interpreted as the expected number of events in an interval of length one, if the waiting times between events were all independent and had exponential distributions with intensity a. Alternatively, the parameter also characterises the probability that an event occurs soon, because $P(U < t) \approx at$ for small t. The exponential distribution with the same mean as the geometric distribution is also shown in Figure 1.15.

The distribution has properties analogous to those of the geometric distribution. Let V be a second exponentially distributed variable with intensity b.

Assume that V is independent of U, and that $t_2 > t_1$, then

$$P(U > t_2 \mid U > t_1) = P(U > t_2 - t_1), \tag{1.14}$$

$$E(U) = \frac{1}{a}, \tag{1.15}$$

$$\text{Var}(U) = \frac{1}{a^2}, \tag{1.16}$$

$$P(U < V) = \frac{a}{a+b}, \tag{1.17}$$

and

$$\min(U, V) \sim \text{Exp}(a + b). \tag{1.18}$$

Property (1.14) restates the Markov property, that all that matters is the present state (and not how long the process has been in that state or which states it resided in previously). The joint properties of U and V are important because waiting for competing, but independent events, such as coalescent events, mutation events, migration events, etc., is at the core of coalescent theory. Waiting for the first of two possible events to occur is again an exponential variable and the probability that one of two types of events is the first to occur only depends on the relative values of the intensities (property (1.17)).

The geometric distribution can be approximated with an exponential distribution in various ways. Here we focus on one particular approximation that is useful in deriving the continuous time coalescent. Let T have a geometric distribution with parameter p, that is,

$$P(T \geq j) = (1 - p)^j. \tag{1.19}$$

If p is small T is typically large and one might measure T on a smaller time scale to counterbalance that p is small. Assume M is some large number such that $a = pM$ and $t = j/M$ are both small compared to M. In the coalescent context M will be of the order of $2N$, p of the order of $1/(2N)$, and j of the order of $2N$. Rewriting $(1 - p)^j$ as

$$\left(1 - \frac{pM}{M}\right)^{M \cdot j/M} = \left(1 - \frac{a}{M}\right)^{tM}, \tag{1.20}$$

we obtain

$$P(T \geq j) = P(T/M \geq t) \approx e^{-at} \tag{1.21}$$

from standard mathematical analysis. In consequence, $U = T/M$ is approximately exponential with intensity a.

This approximation involves a change in how time is measured: Instead of measuring in discrete units, time is measured in continuous time such that one unit of continuous time corresponds roughly to M discrete units. In coalescent theory this will be $2N$ discrete generations. This is accomplished by dividing T by M. To jump between the two time scales we either divide by M or multiply by M. As just mentioned, to go from discrete time units we divide j by M and, conversely, we multiply t by M to obtain discrete time from continuous time: for example, if $t = 2.353$ and $M = 100$ then $j = tM \approx 235$.

1.7 The discrete-time coalescent

With the concept of the Wright–Fisher model of reproduction and the properties of the associated important probability distributions, the binomial distribution, the geometric distribution, and the exponential distribution, at hand, we are in a position to derive the basic coalescent process. The presentation does not follow the original formulation by Kingman (1982a, b) because his formulation is mathematically advanced. What will be derived is what Kingman termed the *n-coalescent* or just the coalescent for a sample of n genes. The coalescent for an infinite (or very large) sample will briefly be discussed at the end of this chapter.

1.7.1 Coalescence of a sample of two genes

What is the distribution of the waiting time until the MRCA of two genes sampled in a haploid model with $2N$ genes? The probability that these two genes find an ancestor in the first generation back in time is $1/(2N)$—the first can choose its parent freely, but the second gene must choose the same parent as the first gene, which is one out of $2N$ possibilities. The probability that the two genes have different ancestors is therefore $1 - 1/(2N)$.

Since sampling in different generations is independent of each other, the probability that two genes find a common ancestor j generations back in time is

$$\left(1 - \frac{1}{2N}\right)^{j-1} \frac{1}{2N}. \tag{1.22}$$

In the first $j - 1$ generations they chose different ancestors, and then in generation j they chose the same ancestor. Thus the coalescence time T_2

for two genes to find a MRCA is distributed as

$$P(T_2 = j) = \left(1 - \frac{1}{2N}\right)^{j-1} \frac{1}{2N},$$

(1.23)

$j = 1, 2, \ldots$, which implies that T_2 is geometrically distributed with parameter $1/(2N)$. The mean of T_2 is therefore $E(T_2) = 1/(1/(2N)) = 2N$ generations. Thus, the expected time until a MRCA is the same as the number of genes in the population. In the example shown in Figure 1.14, where $2N$ is ten, the probability that two randomly picked genes in one generation has the same parent is $1/10$ and the expectation of the waiting time for them to find a common ancestor is ten generations.

1.7.2   Coalescence of a sample of n genes

The waiting time for k ($\leq n$) genes to have less than k ancestral lineages can also be calculated. The probability that k genes have k different ancestors in the previous generation is

$$\frac{(2N-1)}{2N} \frac{(2N-2)}{2N} \cdots \frac{(2N-k+1)}{2N} = \prod_{i=1}^{k-1} \left(1 - \frac{i}{2N}\right)$$

$$= 1 - \sum_{i=1}^{k-1} \frac{j}{2N} + O\left(\frac{1}{N^2}\right) = 1 - \binom{k}{2}\frac{1}{2N} + O\left(\frac{1}{N^2}\right),$$

(1.24)

where $O\left(1/N^2\right)$ is all terms which are divided by N^2 or any higher power of N. Since we assume that n is much smaller than N, $O\left(1/N^2\right)$ is negligible and can be ignored (the effect of this is exemplified below). This approximation is equivalent to ignoring the possibility that more than one pair of genes find a common ancestor in the same generation.

The reasoning behind equation (1.24) is analogous to the two genes case. The first gene can choose freely among the $2N$ genes, the second gene must choose a different parent and can only choose between $2N - 1$ parents. The third gene can only choose among $2N - 2$ possible genes and so forth. Thus, when n is much smaller than N, the probability that no coalescence event occurs is

$$1 - \binom{k}{2}\frac{1}{2N},$$

(1.25)

and the probability of a coalescence event in a given generation is

$$\binom{k}{2}\frac{1}{2N}.$$

(1.26)

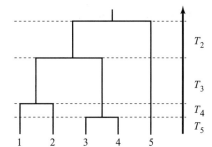

Figure 1.16 The different time epochs in a coalescent tree. T_j, $j = 2, 3, 4, 5$ is the time while there are j ancestors to the sampled five genes.

In consequence the probability that two genes out of the k genes finds a common ancestor $T_k = j$, $j = 1, 2, \ldots$, generations ago is

$$P(T_k = j) \approx \left\{ 1 - \binom{k}{2} \frac{1}{2N} \right\}^{j-1} \binom{k}{2} \frac{1}{2N}, \tag{1.27}$$

and T_k has approximately a geometric distribution with parameter $\binom{k}{2}/(2N)$. The terminology is shown in Figure 1.16. Because all pairs of genes are equally likely to find a common ancestor, the pair that finds a common ancestor is chosen with equal probability among the $\binom{k}{2}$ possible pairs. The times $T_2, \ldots, T_n$ are independent.

1.7.3 Example: Effect of approximations

Relating the quantities of the previous sections to the example of Figure 1.14, the probability that all three genes have one ancestor in the previous generation is $\frac{1}{10} \frac{1}{10} = \frac{1}{100}$, since the first gene can choose a parent freely, while the next two genes must choose the same parent as the first gene. The probability that three genes have three different ancestors is $\frac{10 \cdot 9 \cdot 8}{10 \cdot 10 \cdot 10}$. The remaining possibility that the three genes have two parents in the previous generation (i.e. one pair of the three possible pairs has only one ancestor) is then $1 - \frac{10 \cdot 9 \cdot 8}{1000} - \frac{1}{100} = \frac{27}{100}$. Using the approximate probability it comes out as

$$\binom{3}{2} \frac{1}{10} = \frac{3}{10}. \tag{1.28}$$

There is a slight difference between the true value and the approximate value, since $\frac{27}{100} - \frac{3}{10} = -\frac{3}{100}$. If the approximate value of $\frac{3}{10}$ is used here, the expected number of generations before any of the three genes have common ancestors is $10/3 \approx 3.33$. $2N = 10$ is very small. For $2N > 100$, the agreement is excellent.

The accuracy of the approximation for large N leads to a formulation of the coalescent with two convenient properties: A model that uses continuous time and that additionally is independent of $2N$. To accomplish this we need the exponential distribution that was introduced previously.

1.8   The continuous time coalescent

In the Wright–Fisher model time is measured in discrete units, generations. It is conceptually and computationally advantageous to consider continuous time approximations. A natural choice for the coalescent has been to scale in continuous time, so that one unit of time corresponds to the average time for two genes to find a common ancestor, which was just shown to be $2N$ generations. Again, note that other treatments of the coalescent process prefer scaling time by N (or occasionally $4N$) rather than $2N$, leading to results differing by a factor of two. Using any of these transformations of time, the coalescent becomes independent of the population size. It will only be used if we want to translate time back into generations. This emphasises that the structure of the coalescent process is the same for any population as long as the sample size n is small compared to the population size $2N$; only the time scale differs between populations when $2N$ differs.

To derive the continuous coalescent process we let $t = j/(2N)$, where j is time measured in generations. It follows that $j = 2Nt$ translates continuous time t back into generations j. (If $j = 2Nt$ is not an integer j is truncated to the nearest lower integer. For example, $2Nt = 2 \cdot 10^4 \cdot 1.23241$ is truncated to 24648.) The waiting time, T_k^c, in the continuous representation for k genes to have $k-1$ ancestors is exponentially distributed, $T_k^c \sim \text{Exp}(\binom{k}{2})$, that is,

$$P(T_k^c \le t) = 1 - e^{-\binom{k}{2}t}. \tag{1.29}$$

This is derived from equations (1.21) and (1.27) letting $t = j/(2N)$, $M = 2N$, and $p = \binom{k}{2}/(2N)$. A continuous time realisation of the coalescent is shown in Figure 1.17, with time scaled in generations on the left and in units of $2N$ generations on the right.

It is now possible to describe a stochastic algorithm that samples genealogies for n genes. Only the version for continuous time is given; the discrete version is analogous to the continuous. We refer to the continuous time coalescent as the basic coalescent or the basic coalescent process.

Algorithm 1

1. Start with $k = n$ genes.
2. Simulate the waiting time T_k^c to the next event, $T_k^c \sim \text{Exp}(\binom{k}{2})$.

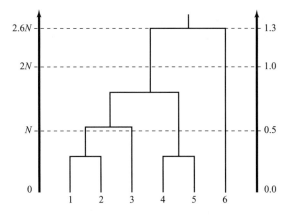

Figure 1.17   A continuous time genealogy with time measured in units of generations (left) and in units of $2N$ generations (right).

3. Choose a random pair (i, j) with $1 \leq i < j \leq k$ uniformly among the $\binom{k}{2}$ possible pairs.

4. Merge i and j into one gene and decrease the sample size by one, $k \rightarrow k - 1$.

5. If $k > 1$ go to 2, otherwise stop.

A computational advantage of going back in time when simulating the history of a present day sample relative to an alternative forward approach is that in the former one only needs to keep track of the ancestry of the sample of interest, while in a forward approach the whole population needs to be traced. Referring back to Figure 1.14, a forward approach would have to calculate all edges in the whole illustration and additionally make sure that sufficiently many generations had been simulated, such that the MRCA to all extant sequences had been found. A backward approach would only have to find the highlighted edges.

From now on we will only refer to the continuous model unless stated otherwise and the superscript c is dropped, thus T_k^c is denoted by T_k. Whenever we think of a particular population with a certain size $2N$, time can be translated back into generations by multiplication by $2N$.

1.9   Calculating simple quantities on a coalescent tree

In the continuous coalescent, the properties of the exponential distribution make it easy to calculate a number of important quantities on the genealogy.

1.9.1   The height of a tree

Consider the coalescent tree depicted in Figure 1.16. The height, H_n, of the tree of a sample of size n is the sum of time epochs, T_j, while there are $j = n, n - 1, \ldots, 2$ ancestors. The distribution of H_n is obtained as a convolution of the exponential variables,

$$P(H_n \le t) = \sum_{k=1}^{n} e^{-\binom{k}{2}t} \frac{(-1)^{k-1}(2k-1)n_{[k]}}{n_{(k)}}, \qquad (1.30)$$

where $n_{[k]} = n(n-1) \cdots (n-k+1)$, and $n_{(k)} = n(n+1) \cdots (n+k-1)$. However, the mean of H_n is easily obtained

$$E(H_n) = \sum_{j=2}^{n} E(T_j) = 2 \sum_{j=2}^{n} \frac{1}{j(j-1)} = 2\left(1 - \frac{1}{n}\right). \qquad (1.31)$$

Similarly, because the T_js are independent, the variance of H_n is

$$\mathrm{Var}(H_n) = \sum_{j=2}^{n} \mathrm{Var}(T_j) = 4 \sum_{j=2}^{n} \frac{1}{j^2(j-1)^2}. \qquad (1.32)$$

The sum in (1.31) grows towards 2 (scaled in $2N$ generations) as the number of genes increases. The expected waiting time for n genes to find their common ancestor is less than twice that of the expected waiting time for two genes to find their common ancestor ($E(H_2) = E(T_2) = 1$). This is a counterintuitive fact for most people when initially introduced to the coalescent. The example also shows that in the continuous time coalescent we can at least formally consider a sample of infinite size because an infinite sample finds a common ancestor in expected time $2(1 - 1/\infty) = 2$, thus in finite time. It is mathematically convenient that the coalescent model applies for even large sample sizes; sample sizes that might be larger than the size ($2N$) of the Wright–Fisher population it seeks to approximate. These considerations connect back to the discussion at the end of section 1.4.2: In a large population it takes only a few generations before the whole population has only a small number of ancestors; the number of generations is much smaller than the size of the population. The variance of H_n goes to a finite value for increasing n: $\frac{4}{3}(\pi^2 - 9) \approx 1.159$. The sum converges very quickly. One property of H_n is a disproportionately large contribution from T_2, the time from two ancestors to one ancestor. Table 1.4 shows the contribution of different epochs to the variance of H_n.

The property that T_2 contributes significantly to the time of the MRCA can also be seen in Figure 1.18. This figure shows that the variation in the height of the coalescent tree for a sample of twenty is large, and that most of it is caused by the variance in T_2.

Table 1.4   Variance of H_n and L_n and the ratio of the variance of H_n, respectively L_n, to that of T_2

n	2	3	4	5	6	10	15	20
$\text{Var}(H_n)$	1.000	1.111	1.139	1.149	1.153	1.158	1.159	1.159
T_2 contribution	1.000	0.900	0.877	0.870	0.867	0.864	0.863	0.863
$\text{Var}(L_n)$	4.000	5.000	5.444	5.694	5.854	6.159	6.304	6.375
T_2 contribution	1.000	0.800	0.734	0.702	0.683	0.649	0.635	0.627

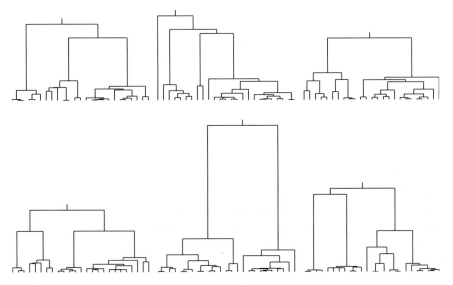

Figure 1.18   A sample of six realisations from the coalescent relating twenty-five genes.

1.9.2   The total branch length of a tree

In contrast to H_n, the distribution of the total branch length L_n has a nice form

$$P(L_n \le t) = (1 - e^{-t/2})^{n-1}. \tag{1.33}$$

The mean of L_n is most easily obtained by weighting the coalescent times by the number of lineages that exist in that epoch,

$$E(L_n) = \sum_{j=2}^{n} j E(T_j) = 2 \sum_{j=1}^{n-1} \frac{1}{j}. \tag{1.34}$$

This sum does not converge for large n, but grows slowly with n. In fact, it is proportional to the natural logarithm of n,

$$\sum_{j=1}^{n-1} \frac{1}{j} \approx \log(n). \tag{1.35}$$

We can use the mean of L_n to get a sense of how much history genes in a sample share. The genes would share the least history if they all sprung from a common ancestor (assuming they had a common ancestor) some time ago and then evolved along distinct lineages. If the time until the common ancestor is the same as in the basic coalescent, namely $E(H_n) = 2(1 - 1/n)$, then the total branch length would be n times that quantity, or $2n(1 - 1/n) = 2(n - 1)$. Comparing this to the mean of L_n in the basic coalescent, we find the ratio

$$\frac{E(L_n)}{2(n-1)} = \frac{\sum_{j=1}^{n-1} \frac{1}{j}}{n-1} \approx \frac{\log(n)}{n-1}. \tag{1.36}$$

This number is small even for small n; $n = 5$: 52%, $n = 10$: 31%, and $n = 100$: 5.2%. For example if $n = 10$ then on average $100\% - 31\% = 69\%$ of the total time in the genealogy of a sample is shared between two or more genes. Thus coevolution of genes is the rule rather than the exception. The variance of L_n is

$$\text{Var}(L_n) = \sum_{j=2}^{n} j^2 \text{Var}(T_j) = 4 \sum_{j=1}^{n-1} \frac{1}{j^2}, \tag{1.37}$$

which converges to $2\pi^2/3 \approx 6.579$ as n increases. This implies that for large n, L_n is narrowly centered around $E(L_n)$. Table 1.4 shows the variance contribution of T_2 to the total variance of L_n.

1.9.3   The effect of sampling more sequences

Sampling more sequences is less effective than most would intuitively think. Figure 1.19 shows an example where a sample of ten sequences (in bold) is embedded in a larger sample of fifty sequences. Most of the deep branches (those nearer the root) are already in the tree with ten sequences. This is not surprising since the expected height of a tree for fifty genes is 1.96, while the expected height for a tree for ten genes is 1.80. Thus, a fivefold increase in genes on the average leads to less than 10% increase in height. The increase in branch length is also slow. In the present case the relative increase is $[E(L_{50}) - E(L_{10})]/E(L_{10}) = 58\%$, which again should be compared to

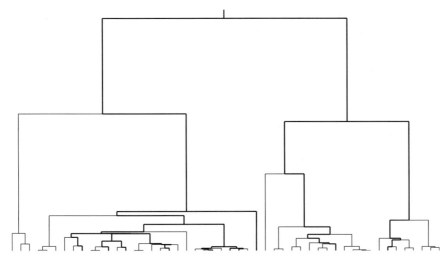

Figure 1.19   The effect of adding more sequences to a sample. A sample of ten sequences (in bold) are extended with forty more sequences. This mainly changes the lower part of the coalescent tree.

a fivefold increase in sample size. The effect of sampling more sequences will be taken up again in more detail in Section 3.4. For example, the probability that the smaller sample shares MRCA with the larger sample is derived.

1.10   The effective population size

As mentioned previously a real physical population is not likely to behave reproductively as the Wright–Fisher model. Most real populations show some form of reproductive structure, either due to geographical proximity of individuals (or genes) or due to social constraints (e.g. all females may not be allowed to marry any male, and vice versa). Also it is likely that the number of descendants of a gene in one generation does not follow the Poisson distribution with intensity one. If the real population has been of constant size over time then the mean number of descendants is bound to be one, but the variance of the distribution might differ from the variance in the Poisson (here one).

When the Wright–Fisher model, or the basic coalescent, is used to model a real physical population, the size $(2N)$ of the population in the (haploid) Wright–Fisher model cannot be taken to be the size of the real population. For example, many human genes had a MRCA less than 200,000 years ago. If we count one generation as 20 years then N should be less than $200,000/(4 \cdot 20) = 2.500$ (the expected TMRCA of the whole population is $4N$), which is unrealistically small for the real population. However, it suggests that the real population might be approximated by a Wright–Fisher model with $N = 2.500$.

For a real population or some model population (e.g. a diploid Wright–Fisher model) the population size of the haploid Wright–Fisher that in some sense best approximates the real population (or the model) is called the effective population size, N_e. Thus, in the example above N_e might be 2.500. There are several ways of defining N_e (e.g. see Ewens 2004). Often different definitions agree or agree approximatively for large values of N_e. Here we shall only be concerned with one definition of N_e and a generalisation of that definition that turns out to be particularly useful for models with variable population sizes (Chapter 4). The measure is called the *inbreeding effective population size*, which Ewens (2004) denoted by $N_e^{(i)}$ to distinguish it from other related quantities. Here it is just denoted by N_e. It is defined by

$$N_e = \frac{1}{2P(T_2 = 1)}, \tag{1.38}$$

where T_2 is in generations. The generalisation is defined by

$$N_e^{(t)} = \frac{E(T_2)}{2} \tag{1.39}$$

where t stands for time and T_2 again is in generations. The main and important difference between the two measures is that N_e is related to the immediate past (the previous generation), whereas $N_e^{(t)}$ is related to the number of generations until a MRCA is found. For the haploid Wright–Fisher model the two definitions agree. Here, $P(T_2 = 1) = 1/(2N)$ and $E(T_2) = 2N$. Thus, $N_e = N$ and $N_e^{(t)} = N$. In the diploid model with N_f females and N_m males ($N_f + N_m = N$),

$$P(T_2 = 1) = \left(1 - \frac{1}{2N}\right) \frac{N}{8N_f N_m} \tag{1.40}$$

and, thus,

$$N_e = N_e^{(t)} \approx \frac{4N_f N_m}{N} = 4c(1 - c)N \tag{1.41}$$

for large N, $N_f = cN$, and $N_m = (1 - c)N$. Again the two definitions agree. In particular, if $N_f = N_m$ ($c = \frac{1}{2}$) then $N_e = N_e^{(t)} = N$. If population size varies with time then the two definitions disagree because N_e depends on the particular generation we focus on.

To give a few other, more extreme, examples, assume that in each generation three genes are chosen and each creates a third of the next generation. The probability that two randomly chosen genes have the same parent would then be a third, independently of population size. Then $N_e = N_e^{(t)} = 1.5$. Or assume that $\sqrt{2N}$ genes are chosen each to have $\sqrt{2N}$ offspring. The probability of having identical parents in the previous generation is then $1/\sqrt{2N}$ and $N_e = N_e^{(t)} = \sqrt{2N}/2$. These are two extreme examples.

Approximating the evolution of a real (or model) population by a haploid Wright–Fisher model with population size $2N = 2N_e$ implies that one particular aspect of the real population is mimicked in the haploid Wright–Fisher model, namely the expected TMRCA of two genes. Other quantities might differ, for example, the variance of the TMRCA, or the TMRCA of a sample of size n, and so on.

In Chapter 4 models with population structure and variable population sizes are introduced and the effective population size provides a standard by which these models can be compared.

1.11   The Moran model

Moran (1958a) proposed an alternative model to the Wright–Fisher model. Moran's model has overlapping generations, in contrast to the Wright–Fisher model that has non-overlapping generations. The population consists of $2N$ haploid individuals or genes. (A diploid version can be formulated as well. Here we focus on the haploid version for convenience.) A new generation is formed from the previous generation by sampling randomly one gene to give birth to a new gene, and one gene to die, see Figure 1.20. The gene that dies is not allowed to be the gene chosen to give birth. (Other formulations of the Moran model allow for the two genes to be one and the same.) All other genes survive to the next generation. The way the Moran model is constructed automatically rules out the possibility of multiple coalescent events in the same generation or that more than two genes share the same common ancestor in the previous generation.

The probability that two genes share a common ancestor in the previous generation is $1/(N(2N-1))$ because one out of the possible $\binom{2N}{2}$ pairs has the desired property. Thus the waiting time until a MRCA of two genes is a geometric distribution with parameter $1/(N(2N-1))$ and the natural time scale is in units of $N(2N-1)$ generations, rather than in units of

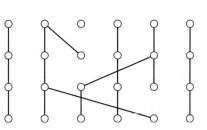

Figure 1.20   The Moran model. Shown is a few overlapping generations in the Moran model. In each generation one gene is chosen to give birth to a new gene and one gene is chosen to die. Because only one gene reproduces in each generation the appropriate time scale is different from that in the Wright–Fisher model.

$2N$ generations. When adjusted for the differences in time scale, the basic coalescent serves as an approximation for both models.

The Moran model is often applied by theoreticians because many calculations turn out to be particularly simple in this model, whereas a similar calculation in the Wright–Fisher model might be highly intractable. However, the Moran model appears to have less appeal to biologists and the Wright–Fisher model is thus more frequently used as a model of population reproduction.

1.12   Robustness of the coalescent

It has already been mentioned that the haploid and the diploid Wright–Fisher models, as well as the Moran model, tend to give probabilistically similar genealogies for large N. This is an example of what is called robustness of the coalescent: The continuous time coalescent can be used as an approximation of many different discrete time population reproduction models. Kingman showed that this is the rule rather than the exception. The effective population size defines the appropriate time scale and provides the transversion factor between discrete time units and continuous time.

Among the 1982 contributions from Kingman, this must be the most practical, in the sense that it legitimises the the use of the coalescent beyond the simple Wright–Fisher based models and the Moran model.

Recommended reading

Ewens, W. J. (2004), *Mathematical Population Genetics*, 2nd edn, Springer Verlag.

Hudson, R. R. (1991), 'Gene genealogies and the coalescent process', *Oxford Surveys in Evolutionary Biology* 7, 1–49.

Kingman, J. F. C. (1982*b*), 'On the genealogy of large populations', *J. Appl. Prob.* **19A**, 27–43.

2 From genealogies to sequences

In the previous chapter we discussed how genes or sequences are related in a population through their common ancestry. However, to model real data a model of how mutations cause changes in the DNA is necessary. To meet this several mutation models have been developed. Historically, the infinite alleles model appeared first (Kimura and Crow 1964), followed by the infinite sites model (Kimura 1969), and the finite sites model (Jukes and Cantor 1969). This development illustrates an increased focus towards analysis of nucleotide sequence data sets, but also a development towards increasing complexity of analysis.

In this chapter we first introduce the three different types of mutation models, then a mathematical framework for working with probabilities of a sample configuration is discussed. We will mainly focus on the infinite alleles and infinite sites models and relate these models to the underlying genealogy of a sample. Mutations are assumed to be selectively neutral. This has the desirable effect that the mutation process can be separated from the genealogical process, because, in the absence of selection, the mutational process and the transmission of genes from one generation to the next are independent processes (genes have the same probability of transmission whatever their type). Thus, a sample configuration for n genes can be simulated using a two step procedure: (1) simulate the genealogy of n genes; (2) add mutations to the genealogy according to the chosen mutation model.

We will make use of forwards and backwards perspectives to calculate and explain quantities of interest. The backwards perspective implies that we are looking backwards in time: what could have happened prior to the present time? The forwards perspective implies that we are looking forwards in time: what could be the next event? We will frequently jump between the two perspectives. In the first two sections, 2.1 and 2.2, the time perspective is with the natural passage of time—from the past towards the present.

2.1 Mathematical models of alleles

2.1.1 The infinite alleles model

This model assumes that the information we have about alleles only allows us to say if they are identical or different. This was conceptually inspired by isozymes, which are differently charged forms of an enzyme. There is no spatial or quantitative information about observed differences. A mutation in an existing allele will always create a new allele not observed before. If we observe alleles, the only information that would be available, would be which alleles are identical. For instance $(1, 4)(2)(3, 5, 7)(6)$ could be such an observation where the notation signifies that four types have been observed, 1 is identical to 4, and 3, 5 and 7 form an identical set. The unique types, 2 and 6 were only observed once. Having observed two different alleles, say 4 and 6, it is impossible to know whether 4 mutated into 6 through a single mutation, or 6 into 4, or whether several mutations separate one of the types from the other.

The actual labels (such as 4 and 6) are without allele specific information and have no biological interpretation. What matters are the sizes of the groups. In the present example that is: two groups of size 1, one of size 2, and one of size 3. This is often abbreviated as $1^2 2^1 3^1$, which is not a product, but a shorthand for the sizes of the sets observed. The superscript denotes the number of groups of a given size.

How this can be coupled to the evolution of a set of genes is described in Figure 2.1. The figure provides an informal and intuitive way of understanding the infinite alleles model. To be able to derive mathematical results for this model (and also the other mutation models to be introduced) we shall return to the Wright–Fisher model in Section 2.2 and explain how the Wright–Fisher model can be extended with a mutation process.

Let us assume that we can determine the types of these genes at any time point in the order they appear. In the beginning there would only be

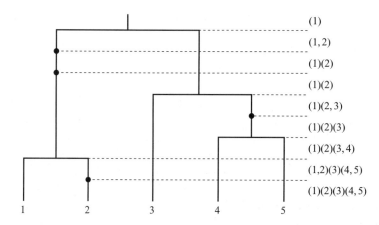

Figure 2.1   Infinite alleles model. Only the set size notation is shown here. The sample's history is followed from the root of the tree towards the present day. The configurations shown to the right of the tree represent the sample configuration as it is between dotted lines. The bullet '•' denotes a mutation event. The labels attached to the genes are those used in the configuration lowest in the figure.

one type (1) or 1^1. The direction of time is here from the past towards the present, from the root towards the present day sample. A split event (a coalescent event when going backwards in time) would then create two identical types, 2^1. A mutation would then modify one of these, creating a different type, leaving two different alleles in one copy each, 1^2. A split event will in the set notation take an allele and duplicate it. In the set size notation, it will remove a set of size k and add a set of size $k + 1$. A mutation event will remove an allele and create a completely new type. In the set size notation, a set of size k will be removed and two sets of sizes 1 and $k − 1$ will be included. (If $k = 1$ the configuration is left unchanged.) In Figure 2.1, $(1)(2, 3)$, or $1^1 2^1$, becomes $(1)(2)(3)$, or 1^3, such that two sets of size one are created and the set of size two is removed. Also $(1)(2)$, or 1^2, becomes $(1)(2, 3)$, or $1^1 2^1$. It is important to understand that the numbers are arbitrary labels and that they do not carry any information about the alleles themselves. For example, $(1)(2)$ remains $(1)(2)$ after the second mutation event counted from the MRCA, and $(1)(2)(3, 4)$ becomes $(1, 2)(3)(4, 5)$ after the third split event.

2.1.2   The infinite sites model

The infinite sites model can be interpreted as describing the evolution of very long DNA strings with low mutation rate at each position. (Therefore, genes are frequently denoted as sequences in this model.) As a consequence a mutation will always happen in a new position. Biologically, the model is justified using the following argument: The number of variable sites in a sample of real sequences is typically small compared to the number of sites which are identical in all sequences. Further, often only one or two nucleotides are found in a given position, indicating that mutations happen rarely. The latter point is of course questionable because mutations in a given position could preferably be between two specific nucleotides or from one specific nucleotide into another.

In the infinite sites model there will always be one or two states in a position of a set of sequences, never more, because each position mutates at most once. All that matters in comparison of two sequences at a given position is whether they are different or not. The two possible alleles are labelled 0 and 1 with no biological meaning attached to the labels. If labels are swapped the interpretation will remain the same. The model also implies that all mutations that have happened in the history of a sample are recoverable, in contrast to the infinite alleles model, where two consecutive mutations in one lineage will only be registered as one overall change. The position of a mutation is random along the sequence.

It is convenient to represent a sequence as a series of zeros and ones, leaving out all positions that are not segregating in the sample. Attach to each segregating site its position in the sequence. An example is given in

Figure 2.2 Representation of infinite sites data. Shown is a sample of size five with four segregating sites. The top row shows the positions, relative to the length of the sequence, as decimal numbers. All positions that are invariable are left out. A history of this sample is shown in Figure 2.3.

0.073	0.294	0.550	0.894
1	1	0	0
1	1	0	1
0	0	0	0
0	0	1	0
0	0	1	0

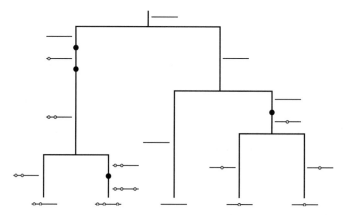

Figure 2.3 Infinite sites model. The sequence is represented by a continuous string with mutations attached. In the example the circle denotes the derived state. However this information is only available if we know the ancestral states of all positions, that is if we know the root sequence. If another combination of states was used as root sequence the tree might not be compatible with the sequence data. All mutations are visible in the present day sequences.

Figure 2.2. Figure 2.3 illustrates an evolutionary history of the sample in Figure 2.2 using a different, but commonly applied, representation of a sequence. The infinite sites model is also discussed in the context of the Wright–Fisher model later in this chapter and some consequences of the infinite sites model will be derived formally.

A series of points should be noted at this stage. A mutation always partitions the set of sequences into two groups. These two groups correspond to an edge on the true tree relating the sequences, in the sense that if this edge was removed the tree would be cut into two trees where all leaves on the first tree shared one character state, while leaves on the other tree shared another character state. In other words, a mutation induces a bipartition of the sample. If the mutation rate was sufficiently high, the true unrooted tree topology of the sequences can be inferred because mutations will be dense on every branch and all sequences in the sample will thus have their own distinctive patterns of mutations. The number of mutations in the two histories in Figures 2.1 and 2.3 are the same, but in the infinite sites example, it is apparent that four mutations have occurred, while in the infinite alleles example, the two mutations on the same branch will only be visible as one

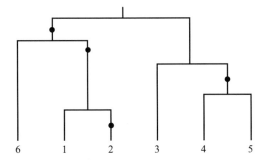

Figure 2.4 Sampling and the infinite alleles model. After sampling of allele 6 both mutations on the branch connecting the ancestor of 2 and 3 with the MRCA become visible in the sample.

mutation, unless sequences were sampled that contained the first but not the second of the two mutations (as in Figure 2.4).

2.1.3 Finite sites model

Modelling has moved towards increased realism. A completely realistic model of DNA evolution would imply a complete specification of the DNA sequence. In population studies such a sequence will have a fixed length and will be assumed to evolve only by mutations. In principle a model should contain a process describing insertions and deletions, but these occur so rarely in population data that sequences normally can be unambiguously aligned, even if insertions/deletions have occurred. One might then choose to discard insertions and deletions.

In the example of Figure 2.5, an original sequence of length eight encountered mutations over time as it evolved. In this series of events, two mutations happened in the same position (position 7), which would be prohibited under the infinite sites model. So four mutations only created three segregating sites in this case. Mutations happening in the same position (but not necessarily in the same lineage) are called recurrent mutations. They create two kinds of partitions that are not possible under the infinite sites model: (1) partitions based on three or four types, as induced by the seventh position in Figure 2.5; (2) bipartitions that do not correspond to a partition defined by a branch in the tree, because of two (or more) mutations in the same direction (e.g. $T \rightarrow A$ twice) or in opposite directions (e.g. $T \rightarrow A$ and $A \rightarrow T$ in different branches). In addition there is the possibility of 'back-mutation', a mutation that erases the effect of the first mutation, for example, $T \rightarrow A$ and subsequently $A \rightarrow T$ in the same branch.

The simplest finite sites model is the Jukes–Cantor model, named JC69, introduced by Jukes and Cantor in 1969: All positions are equally likely to mutate and the mutant is chosen with equal probability among the three

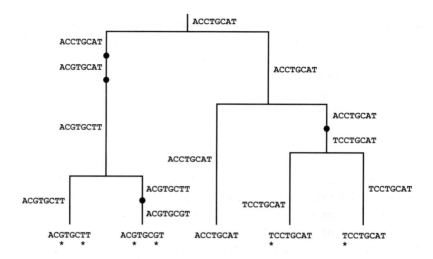

Figure 2.5 Finite sites model. Each nucleotide position might experience more than one mutation. Two sequences can be identical even if several mutations have occurred in their history: Forth and back mutations can erase the stamp of mutations, for example, if A → T at some point back in time and T → A in the same position but closer to the present time.

possible nucleotides. This model was later modified by Kimura (1980) to accommodate the fact that transition events (A ↔ T and C ↔ G) occur at a faster rate than transversion events (all other events). It is called K80. Both JC69 and K80 are unrealistic in the sense that all nucleotides are expected to occur with the same frequency (viz. 0.25) in a random sequence, which is not likely to be the case for any sequence. Felsenstein (1981) modified JC69 to take into account unequal base frequencies (F81), and Hasagawa et al. (1985) combined F81 and K80 to allow for unequal base frequencies and transition/transversion bias at the same time. Still other more sophisticated models have been introduced to account for subtle differences in substitution rates.

An important assumption of the finite sites models discussed above is that positions evolve independently of each other. (This is also an assumption of the infinite sites model because a mutation in one position does not influence the chance of a mutation in another position.) It is possible to model coupled or dependent evolution between neighbouring positions. Such models become increasingly difficult to analyse though they might be more appropriate for the analysis of many real sequences. One special kind of such models are codon models. The biological fact that DNA sequences are translated into proteins creates an additional level of complexity. Each triplet of nucleotides (a *codon*, 64 in total) codes for a specific amino acid that differs from other amino acids in respect of chemical and physical properties. When coding DNA evolves, a substitution might or might not change the encoded amino acid. An amino acid change might have

functional or phenotypical consequences and thereby be of severe importance for the cell's or organism's ability to function and survive. Ideally, a
model of DNA sequence evolution should incorporate information about
the encoded amino acids. However, a population data set rarely shows more
than a few differing amino acids and it becomes practically impossible to
perform inferences in such models because of the huge number of model
parameters compared to the sparse data.

In the infinite alleles model and the infinite sites model the mutation rate
can be taken as an overall rate for all types of events that change the type of a
sequence: Thus the rate can be taken to comprise both the rate of nucleotide
mutations and the rate of insertions and deletions. All that matters is that a
mutation, insertion, or deletion introduces a change in the type of the gene
and that this change is unique to the sample. In contrast, only nucleotide
substitutions are modelled in the finite sites model.

2.2 The Wright–Fisher model with mutation

The Wright–Fisher model was introduced in Chapter 1 as a purely reproductive model without any information about genetic type. We can impose
a process of mutation on top of the process of reproduction in the following
simple way: Each gene chosen to be passed on is subject to a mutation event
with probability u. That is to say, with probability $1 - u$ the gene is copied
without modification (or error) to the offspring and with probability u a
mutation occurs. Under the infinite alleles model a new type not previously
seen in the population is introduced, under the infinite sites model a position along the gene is randomly chosen and the type of that position is
changed (from zero to one, or vice versa), and under the finite sites model
a site is chosen randomly and a mutation occurs in that position according
to a given model. Thus, the overall structure of the mutation process is the
same irrespective of whether an infinite alleles model, an infinite sites model
or a finite sites model is assumed.

In Figure 2.6 three generations are shown. All mutations are selectively
neutral in the sense that the type of the parent does not influence the

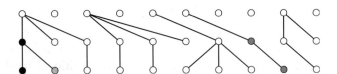

Figure 2.6 The Wright–Fisher model with mutation. Shown are three generations of the
Wright–Fisher model with mutation. In the second generation, counted from the top row,
two descendants of the previous generation are mutated copies of their parent gene. One
descendant of the leftmost mutant in generation two is again a mutated copy of its parent.

probability that a mutation occurs or that the offspring survives. If we follow a lineage from the present time into the past (e.g. the lineage defined by the leftmost gene in Figure 2.6) then there is a chance, u, that the type of the parental gene in generation t differs from the type of the offspring gene in generation $t + 1$. Consequently, the probability that a lineage experiences the first mutation j generations into the past, counting from the present time, is

$$P(T_M = j) = u(1 - u)^{j-1}. \tag{2.1}$$

Here T_M denotes the number of generations until the first mutation event. It follows that T_M is a geometric variable with parameter u. If time is measured in units of $2N$ generations (as in the basic coalescent process) then

$$P(T_M \le j) = 1 - (1 - u)^j \approx 1 - e^{-\theta t/2} = P(T_M^c \le t), \tag{2.2}$$

for large $2N$, where $t = j/(2N)$, $\theta = 4Nu$, and T_M^c denotes time in units of $2N$ generations. The parameter θ is called the population mutation rate or the scaled mutation rate. There are no fundamental reasons why θ is defined as $4Nu$ rather than $2Nu$, but some formulas turn out to simplify when scaled in $4N$, rather than in $2N$. The parameter θ can be interpreted as the expected number of mutations separating a sample of two sequences, since the expected coalescence time for two sequences is $2N$, and thus $2Nu$ mutations are expected on each branch. It transpires that T_M^c is an exponential variable with parameter $\theta/2$. Using T_M^c as a continuous time approximation to T_M is accurate as long as N is large. Henceforth we focus exclusively on the continuous time approximation. (Superscript c in T_M^c is dropped when ambiguities are not possible.)

If we instead consider n disjoint lineages then the time until the first mutation event in any of the n lineages is exponential with parameter $n\theta/2$, that is, n times that of a single lineage because lineages evolve independently of each other. Whether there is a mutation in one gene in one generation is independent of whether there is another mutation in another gene in the same generation. The form of the parameter $n\theta/2$ is now a consequence of property (1.18).

If we wait for mutation events and coalescence events then the parameter of the exponentially distributed waiting time is the sum of the two parameters. Again this is a consequence of property (1.18), because the two types of events are independent of each other. It follows that the parameter is

$$\binom{n}{2} + \frac{n\theta}{2} = \frac{n(n - 1 + \theta)}{2}. \tag{2.3}$$

Whether the first event is a coalescence event or a mutation event is determined by tossing a biased coin. With probability

$$\frac{\binom{n}{2}}{\binom{n}{2} + \frac{n\theta}{2}} = \frac{n - 1}{n - 1 + \theta}, \tag{2.4}$$

the event is a coalescence event and with probability

$$1 - \frac{n - 1}{n - 1 + \theta} = \frac{\theta}{n - 1 + \theta}, \tag{2.5}$$

the event is a mutation event. The pair that merges is chosen randomly among all pairs and the lineage undergoing mutation is chosen randomly among the n lineages.

2.3   Algorithms for simulating sequence evolution

This section deals with simulation of sample histories and simulation of sample configurations, two closely related subjects. Figures 2.4 and 2.5 provide examples of sample histories. A sample history is a sequence of dated events that created the sampled genes from a gene at the root. The sample configuration is obtained by discarding the history, keeping the configuration at the time of sampling. Naturally, if we can simulate sample histories, we can also simulate sample configurations.

Simulation of sample configurations and sample histories are of importance for many reasons. First of all, it provides means to study variation in random samples by repeated generation of sample configurations. Many quantities of interest and distributions of random variables cannot be found explicitly and we must resort to simulation. Second, it gives intuition into the dynamics of the coalescent and the mutation processes. And finally, efficient sampling of sample histories becomes a major issue for inferential procedures (Chapter 6).

Equations (2.3)–(2.5) form the basis of the first algorithm for simulating a set of genes (or sequences) with mutations.

Algorithm 2

1. Put $k = n$, where n is the sample size.
2. Choose an exponential variable with parameter $k(k - 1 + \theta)/2$.
3. With probability $(k - 1)/(k - 1 + \theta)$ the event is a coalescence event and with probability $\theta/(k - 1 + \theta)$ it is a mutation event.

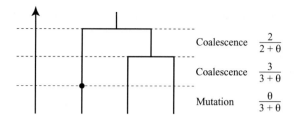

Figure 2.7   Algorithm 2. Simulation of a sample history for $n = 4$ genes. The first event is a coalescence event with probability $(n - 1)/(n - 1 + \theta) = 3/(3 + \theta)$ and a mutation event with probability $\theta/(3 + \theta)$. In this example the event was a mutation event. The algorithm continues until there is only one lineage in the sample. To the right is shown the type of the event and the probability that it occurred.

4. If a coalescence event occurs choose a pair randomly to coalesce. Update $k, k \to k - 1$.

5. If a mutation event occurs choose a lineage to mutate. Leave k unchanged.

6. Continue until k is one.

The algorithm is an extension of Algorithm 1 given in Chapter 1. It is illustrated in Figure 2.7. To determine the types of the present day genes (or sequences) start at the MRCA and move forward in time. Each time a mutation event is encountered the type of the gene is modified according to the chosen mutation model, as illustrated in Figures 2.4 and 2.5. Each time a split event is encountered copy the parent gene onto both descending lineages.

The second algorithm derives from the fact that the waiting time until a mutation occurs along a lineage is exponential with parameter $\theta/2$. This is mathematically equivalent to the following. Consider a branch of length t. The number, M_t, of mutations on the branch is Poisson distributed with intensity $t\theta/2$, that is,

$$P(M_t = j) = \frac{(t\theta)^j}{j!2^j} e^{-t\theta/2}. \tag{2.6}$$

In particular the mean of equation (2.6) is $t\theta/2$. Given the number of mutations on a branch the time of each mutation is random. The numbers and times of mutations on different branches are independent of each other such that branches can be treated one at a time. The process placing mutations on the branches according to equation (2.6) is called a Poisson process.

Algorithm 3

1. Simulate the genealogy of n sequences according to the coalescent process with rate $\binom{k}{2}$ while there are k lineages, that is, following Algorithm 1.

2. For each branch draw a number, M_t, from a Poisson distribution with intensity $t\theta/2$, where t is the length of the branch.

3. For each branch the times of the M_t mutation events are chosen randomly on the branch.

Algorithm 3 uses explicitly that mutations can be tossed onto the genealogy after the genealogy has been simulated. This is in fact an extremely useful property because it makes it easy to generalise Algorithm 3 to other scenarios by changing the way the genealogy is constructed (see Chapter 4). As long as the mutation process is neutral, mutations can be simulated according to the Poisson process and added to the genealogy in a second step. Figure 2.8 illustrates the algorithm.

If we are not interested in the sample histories as such, but just the configuration of the sample, Algorithm 3 can be simplified. Instead of simulating all mutation events on a particular branch (of length t) we can simulate the allelic state of the gene at the end of the branch given the allelic state of the gene at the beginning of the branch. Under the infinite alleles model this amounts to calculating the probability of at least one mutation over time t which is one minus the probability of no mutations, that is, $1 - e^{-t\theta/2}$, and simulate a binary random variable: With probability $1 - e^{-t\theta/2}$ the gene is mutated, and with probability $e^{-t\theta/2}$ the type of the gene is left unchanged.

Under the infinite sites model we simulate a Poisson number of mutations, M_t, and add them to the sequence at the end of the branch. The position of a mutation is chosen randomly along the sequence. (Note that for a given branch of length t with M_t mutations the time of an event is chosen randomly on the branch and the position of a mutation is chosen randomly on the length of the sequence.)

Finally under the finite sites model we focus first on a single nucleotide, instead of the whole sequence. The rate of mutations for a single nucleotide

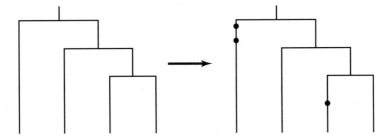

Figure 2.8 Algorithm 3. Two step simulation of sample histories. First, the genealogy of the sample is simulated. Second, mutations are put onto each branch according to a Poisson process with intensity $\theta t/2$, where t is the length of the branch. Mutations are randomly spaced on the branch. The sample configuration is determined by starting at the root moving forward in time and modifying the type of the genes as mutations are encountered.

is $\theta_0 = \theta/L$ if there are L nucleotides. The probability that a nucleotide is in state j given it is in state i, t time units in the past, is

$$P_{ij}(t) = 0.25 \left\{ 1 - e^{-2t\theta_0/3} \right\}, \tag{2.7}$$

if $i \neq j$, where $i, j \in \{A, G, T, C\}$, and

$$P_{ii}(t) = 0.25 \left\{ 1 + 3e^{-2t\theta_0/3} \right\}, \tag{2.8}$$

(see also Figure 2.9). Here we have assumed a Jukes–Cantor model such that each time a mutation happens the new nucleotide is equally likely to be any of the other three nucleotides.

If a sequence is L nucleotides long, the probability that the sequence evolves into some other sequence after time t is obtained as the product of L terms, one for each nucleotide. In particular, the probability that two sequences differ in d positions has probability

$$P(D_t = d) = \binom{L}{d} \left(0.75 \left\{ 1 - e^{-2t\theta_0/3} \right\} \right)^d \left(0.25 \left\{ 1 + 3e^{-2t\theta_0/3} \right\} \right)^{L-d},$$
$$\tag{2.9}$$

where D_t is the number of differences after time t and the binomial coefficient is the number of ways d nucleotides can be chosen out of L possible. The positions of the d differences are chosen randomly among the L possible positions.

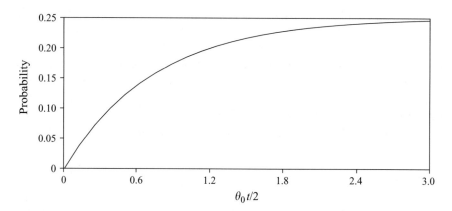

Figure 2.9   Finite sites model. The probability that a nucleotide has changed to a specific different nucleotide as a function of $\theta_0 t/2$. This probability starts as being 0, but then converges from below to 0.25 as $\theta_0 t/2$ becomes large.

2.4   The probability of a sample configuration

In the previous section we have shown how sample configurations of genes can be simulated for three different mutation models. All of these share an underlying common structure, the Poisson process, and they differ only in how mutations modify the genes.

The problem of calculating the probability of data for the infinite alleles, infinite sites and finite sites models are quite different. For infinite alleles the solution was first presented in a seminal paper by Warren Ewens in 1972 and allows the calculation for any data set. For the infinite sites model, it was solved by Griffiths, Ethier, and Tavaré in a series of technical papers in the period 1987–1995. The solutions are calculated via a set of recursions which allows the calculation of the probability for a moderately sized data set. Moderately sized implies here that the number of segregating sites plus number of haplotypes is less than thirty.

For the finite sites model, the problem becomes computationally harder, but also much more useful. If the phylogeny was known, there is an algorithm (Felsenstein 1981) that calculates the probability of a set of nucleotides at the leaves for a fixed phylogeny and mutational process. Since evolution was assumed to occur independently in each nucleotide alignment, the probability for the complete sequences will just be the product of the probabilities for individual nucleotide alignments (this has already been discussed in the previous section, cf. equation (2.9)), that is,

$$P(\mathrm{Seq}_1,\ldots,\mathrm{Seq}_n \mid \mathcal{T},\mathcal{W}) = \prod_{j=1}^{L} P(\mathrm{Nuc}_j \mid \mathcal{T},\mathcal{W}), \qquad (2.10)$$

where Seq_i is the ith sequence, Nuc_j is the jth column of nucleotides, L the number of nucleotides, $\mathcal{T}$ is the topology relating the sequences and $\mathcal{W}$ the set of branch lengths. What is the probability of the data observed in Figure 2.5 ? In this case we do not know the history behind the five sequences, but only the sequences at the leaves—not the tree, no times, and no ancestral sequences. If the tree is described by the coalescent process the probability of the sample, not conditioned on a particular tree, is $P(\mathrm{Seq}_1,\ldots,\mathrm{Seq}_n \mid \mathcal{T},\mathcal{W})$ integrated over all branch lengths and possible topologies,

$$P(\mathrm{Seq}_1,\ldots,\mathrm{Seq}_n) = \int_{\mathcal{T},\mathcal{W}} P(\mathrm{Seq}_1,\ldots,\mathrm{Seq}_n \mid \mathcal{T},\mathcal{W}) f(\mathcal{T},\mathcal{W}) \, d(\mathcal{T},\mathcal{W}),$$

$$(2.11)$$

where $f(\mathcal{T},\mathcal{W})$ is the probability density of $(\mathcal{T},\mathcal{W})$. This integration will use the probability measure defined by the coalescent process and involves two components. First, summing over all coalescent topologies and second integrating over the waiting times corresponding to epochs with different numbers of ancestors. Since the coalescent topologies quickly grow to astronomical numbers and the integral will be high-dimensional, it is not possible to do this exactly for realistically sized data.

2.4.1   Infinite alleles model

Ewens (1972) found the formula for the probability of a sample configuration under the infinite alleles model. The formula can be derived in a number of different ways—here we will show how the probability of a sample configuration fulfils a certain recursion equation. The formula can subsequently be obtained from the recursion. Recursion equations or just recursions are very useful in coalescent theory and we will see many instances of how recursions are used to solve and simplify problems.

   First we introduce some notation. A very similar notation will be used in the next subsection to handle the infinite sites model. Let a_i denote the number of allele classes with i members, $i = 1, \ldots, n$, and define $\mathbf{a} = (a_1, \ldots, a_n)$. Then the total number of genes is

$$n = \sum_{i=1}^{n} i a_i. \tag{2.12}$$

In the example in Figure 2.1, $a_1 = 3, a_2 = 1$, and $a_3 = a_4 = a_5 = 0$ such that $\sum_{i=1}^{5} i a_i = 1 \cdot 3 + 2 \cdot 1 + 3 \cdot 0 + 4 \cdot 0 + 5 \cdot 0 = 5$ as it should be. Further, let $\mathbf{e}_j = (0, \ldots, 0, 1, 0 \ldots, 0)$ with 1 in the jth coordinate entry.

   Starting with a sample of size n the first event (going backwards) can either be a coalescent event or a mutation event. This leads to the recursion

$$P_n(\mathbf{a}) = \frac{\theta}{n - 1 + \theta} P_{n-1}(\mathbf{a} - \mathbf{e}_1)$$

$$+ \frac{n - 1}{n - 1 + \theta} \sum_{a_{j+1} > 0} \frac{j(a_j + 1)}{n - 1} P_{n-1}(\mathbf{a} - \mathbf{e}_{j+1} + \mathbf{e}_j), \tag{2.13}$$

with boundary or initial condition

$$P_1(1) = 1, \tag{2.14}$$

because a sample of size 1 always has one type only. Here subscript n in P_n is used to emphasise sample size.

   The first event is a coalescent event with probability $(n - 1)/(n - 1 + \theta)$ (see Algorithm 2). In that case one class of alleles is chosen among the classes with at least two alleles (only identical types can coalesce), say one of the

a_{j+1} classes with $j + 1$ alleles are chosen. After the coalescent event the number of classes with $j + 1$ alleles is reduced by one and there is one more class with j alleles. That accounts for the new sample configuration after the event. To account for the factor in front of the probability note that the chance that one of the $a_j + 1$ classes with j alleles is chosen to duplicate is $j(a_j + 1)/(n - 1)$ (now moving forwards in time). Thus the probability of a new sampling configuration is weighted by the probability that the new configuration leads to the current configuration given a split event occurs.

Similarly the first event is a mutation event with probability $\theta/(n - 1 + \theta)$. In that case a singleton is removed and the new sampling configuration without the mutated gene is as stated in the first term on the right side. The probability that the new configuration leads to the current configuration is one, because the other genes in the configuration are assumed not to mutate. This is a very common way to derive recursion equations in coalescent theory. The idea is first to ask what could be the most recent event (going backwards in time) and then to ask, for each of the possible new configurations, what is the probability that this configuration leads to the current configuration while going forwards in time. This kind of recursion will be taken up in much more detail in the next subsection.

It allows us to find the probability of any configuration by repeated application of equation (2.13) because the left side is written in terms of probabilities of samples of size $n - 1$. Eventually samples of size one are reached and these have probability one. For example, consider the configuration $a = (a_1, a_2, a_3) = (1, 1, 0)$ with sample size $n = \sum_{i=1}^{3} ia_1 = 3$. It has probability

$$P_3(1, 1, 0) = \frac{\theta}{2 + \theta} P_2(0, 1) + \frac{2}{2 + \theta} P_2(2, 0). \qquad (2.15)$$

The first probability on the right side can be written

$$P_2(0, 1) = \frac{1}{1 + \theta} P_1(1) = \frac{1}{1 + \theta}, \qquad (2.16)$$

and the second term

$$P_2(2, 0) = \frac{\theta}{1 + \theta} P_1(1) = \frac{\theta}{1 + \theta}. \qquad (2.17)$$

Substituting equations (2.16) and (2.17) into (2.15) yields

$$P_3(1, 1, 0) = \frac{\theta}{2 + \theta} \frac{1}{1 + \theta} + \frac{2}{2 + \theta} \frac{\theta}{1 + \theta} = \frac{3\theta}{(2 + \theta)(1 + \theta)}. \qquad (2.18)$$

However, as mentioned previously, an exact solution to the recursion (2.13) can also be found. It is known as Ewens' sampling formula and is given by

$$P_n(a_1, a_2, \ldots, a_n) = \frac{n!}{\theta_{(n)}} \prod_{j=1}^{n} \left(\frac{\theta}{j} \right)^{a_j} \frac{1}{a_j!}. \tag{2.19}$$

where $\theta_{(n)} = \theta(\theta+1) \cdots (\theta+n-1)$. Thus, equation (2.18) could be derived directly from equation (2.19). Ewens' sampling formula places greatest probability on configurations with a few common types and others very infrequent. Table 2.1 provides an example.

Table 2.1   Ewens' sampling formula[a]

Configuration	Probability	K_n	$P(K_n = k)$
$1^1 7^1$	0.143	2	0.324
8^1	0.125	1	0.125
$1^1 2^1 5^1$	0.100	3	0.325
$2^1 6^1$	0.083	2	
$1^2 6^1$	0.083	3	
$1^1 3^1 4^1$	0.083	3	
$3^1 5^1$	0.067	2	
$1^2 2^1 4^1$	0.063	4	0.169
$1^1 2^2 3^1$	0.042	4	
$1^3 5^1$	0.033	4	
4^2	0.031	2	
$2^2 4^1$	0.031	3	
$2^1 3^2$	0.027	3	
$1^2 3^2$	0.027	4	
$1^3 2^1 3^1$	0.027	5	0.048
$1^4 4^1$	0.010	5	
$1^2 2^3$	0.010	5	
$1^4 2^2$	0.005	6	0.008
$1^5 3^1$	0.003	6	
2^4	0.003	4	
$1^6 2^1$	0.001	7	0.001
1^8	0.000	8	0.000
Total	1.000		1.000

[a] The probability of the twenty-two different sampling configurations of a sample of eight genes, shown in set notation for convenience and ordered after highest probability. The scaled mutation rate is $\theta = 1$. A configuration $(a_1, a_2, \ldots, a_8)$ corresponds to $1^{a_1} 2^{a_2} \cdots 8^{a_8}$ in the set notation.

The probability of the number, K_n, of different allele classes in the sample can also be derived. Note that

$$K_n = \sum_{j=1}^{n} a_j, \qquad (2.20)$$

such that the probability that $K_n = k$ can be found from equation (2.19) by summing over all possible configuration with k types. It can also be derived from a recursion very similar to, but less complex than equation (2.13),

$$P(K_n = k) = \frac{\theta}{n - 1 + \theta} P(K_{n-1} = k - 1) + \frac{n - 1}{n - 1 + \theta} P(K_{n-1} = k),$$

$$(2.21)$$

and initial condition $P(K_1 = 1) = 1$. It has solution

$$P(K_n = k) = S(n, k) \frac{\theta^k}{\theta_{(n)}}, \qquad (2.22)$$

where $S(n, k)$ is the Stirling number of the first kind, depending on n and k only. The exact form of $S(n, k)$ does not matter here. Note that the conditional probability of a sample configuration given $K_n = k$ does not depend on θ,

$$P(a_1, a_2, \ldots, a_n \mid K_n = k) = \frac{n!}{S(n, k)} \prod_{j=1}^{n} \frac{1}{j^{a_j} a_j!}. \qquad (2.23)$$

It is an interesting quantity because it shows that K_n carries most of the information in the sample about θ. (Note that the likelihood (2.19) is the product of (2.22) and (2.23)). One can argue that if $K_n = k$ is known then the configuration a does not provide any further information about θ than already is contained in $K_n = k$. As an example if $a_1 = (a_1, a_2, a_3) = (0, 2, 0)$ and $a_2 = (1, 0, 1)$ then $K_n = 2$ in both cases and we cannot learn anything about θ from a_1 that cannot be learned from a_2 and vice versa.

Several quantities derived from the distribution of K_n are of interest. The mean and the variance of K_n are respectively

$$E(K_n) = \sum_{j=0}^{n-1} \frac{\theta}{j + \theta}, \qquad (2.24)$$

and

$$\text{Var}(K_n) = \sum_{j=1}^{n-1} \frac{\theta j}{(j+\theta)^2}. \tag{2.25}$$

Also the number of singletons (a_1) is of interest. The mean of a_1 is

$$E(a_1) = \frac{n\theta}{n-1+\theta}, \tag{2.26}$$

which is approximately θ for large n. We will return to these quantities again in Section 2.5.

2.4.2 Infinite sites model

Fewer quantities can be calculated under the infinite sites model than the infinite alleles model. However, it is still possible to derive recursions that for a smallish data set can be solved numerically or even analytically. For realistically sized data sets it is impossible to provide the solution analytically and in many cases also impossible to provide reliable solutions by simulation.

Let $\mathbf{n} = (n_1, n_2, \ldots, n_k)$ be a sample configuration taken from the infinite sites model. Here n_i is the number of genes of a given type i such that $\sum_{i=1}^{k} n_i = n$. Each type is determined by a string of unordered mutations. Because recombination is assumed absent the positions of mutations can be ignored. Although the ancestral state of the sample is generally unknown we will here for simplicity assume it is known. Thus for each mutation it is known whether a given type i carries the mutant or the ancestral type. The probability of a sample configuration can be decomposed according to the configuration of the sample after the first event going backwards in time. This event can either be a coalescent event or a mutation event. Because of the infinite sites assumption only singleton alleles can be removed by mutation. This leads to the recursion

$$P_n(\mathbf{n}) = \frac{n-1}{n-1+\theta} \sum_{n_j>1} \frac{n_j-1}{n-1} P_{n-1}(\mathbf{n} - \mathbf{e}_j)$$

$$+ \frac{\theta}{n-1+\theta} \sum_{\text{singletons}} \frac{n_i+1-\delta_{ij}}{n} P_n(\mathbf{n} - \mathbf{e}_j + \mathbf{e}_i), \tag{2.27}$$

with initial condition $P_1(1) = 1$ and where the sum in the last term is over all singleton types, i.e. types with multiplicity $n_j = 1$. The number δ_{ij} is one if $i = j$ and otherwise it is 0, and $\mathbf{e}_j = (0, \ldots, 0, 1, 0, \ldots, 0)$ is a vector of length k (the number of types) with the jth entry equal to one.

If a mutation is removed the new type can be identical to another type in the sample or be different from all other types. In the latter case the new type is denoted j (i is put equal to j), thereby reusing the label of the type that no longer exists. Defining the degree of the recursion as the number of sequences plus the number of segregating sites, the degree of the terms on the right side is one less than the degree of the terms on the left side. Thus repeated applications of equation (2.27) eventually lead to a configuration with one sequence and no segregating sites.

Again we use a backward–forward approach to argue for the correctness of equation (2.27). If the first event is a coalescence event then there will be one less of some type, say type j. The probability that one of the genes of type j is chosen to duplicate is $(n_j - 1)/(n - 1)$ which accounts for the factor in front of the probability in the first term on the right side of equation (2.27). If the first event is a mutation event then the chance that the new sample configuration leads to the current configuration is $(n_j - 1 + \delta_{ij})/(n - 1)$, which is the factor in front of the second term.

In Figure 2.10 the possible histories are illustrated on a data example with five sequences and four mutations. The ancestral state is assumed known (referred to as 0) and the derived states (mutants—referred to as 1) are indicated as little balls. Had another ancestral allele than 0000 been chosen, 0s could switch to 1s, when tracing the history going back in time. Since there are five sequences and four mutations, and the root configuration is a single mutationless sequence; any path from the present configuration to the root will have four edges representing coalescences and four edges representing mutations. The illustration has all possible ancestral configurations as nodes. A configuration that can go to another configuration (back in time), defines an edge pointing from the first to the last. The graph has a sink (all nodes have a path leading to it) in the root, whose state is known and has probability one. This graph also only has one source (start at any node and walk against the arrow and this node will be reached), namely the data configuration.

The number of possible distinct ways to generate the present configuration from the ancestral allele can easily be found by a dynamical algorithm that keeps track of how many paths there are from the root to any configuration. In Figure 2.10 there are seventy-two possible paths from source to sink. Mutations are labelled by their position in the sequence whereas sequences are unlabelled. Labelling of sequences is immaterial to the involved probabilities. However, the position of a mutation can be moved without changing the probability of the data. (This is only true as long as recombination is not allowed.)

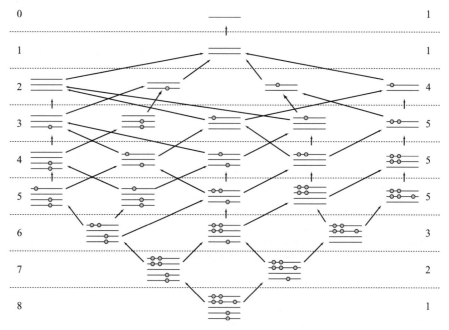

Figure 2.10 The observed data set consists of five sequences that have four varying sites. It is known that the balls are the derived states and going back in time they will disappear. Since they appear one by one and coalescent events only involve two identical sequences, a history of a sample must consist of a path involving four mutations and four coalescent events. However, the order of these and which sequences and mutations are involved is not normally unambiguously determined by the data. For this data set there are seventy-two possible orders of events from data set to configurations. To make this diagram is straightforward. Start with the data set. Which events could have created this? A coalescent of the two bottom sequences or the unique mutation in the third sequence. Collapsing the two sequences or removing the mutation would create two different configurations with four and four sequences respectively. Both of these configurations will have an arrow pointed to it from the original data configuration. This procedure of defining possible ancestors with arrows pointing to them from the configuration they could create (forward in time) by a mutation or a duplication will eventually lead to the root with no mutants and no possibility for coalescence. The left row of numbers indicates how many events have happened relative to the ancestral sequence. The right row of numbers is the number of sequence configurations, corresponding to a given number of events counted from the ancestral sequence. In this diagram mutations are ordered by their position.

In passing from the diagram in Figure 2.10 to the diagram of Figure 2.13, the order of the mutations has been ignored. The unordered configuration corresponds to a rooted gene tree. The probability of the sample configuration can be obtained from the probability of the gene tree dividing by 2 because this is the number of ways the two leftmost mutations can be ordered (see Figure 2.11).

The central argument in deriving rates and probabilities on these graphs is illustrated in Figure 2.12. The simplicity of the history of the example in Figure 2.10 hinges completely on the infinite sites assumption and so

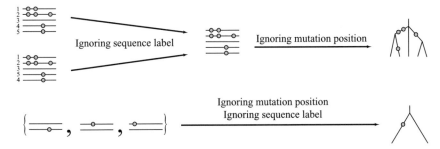

Figure 2.11 The data set illustrated in Figure 2.10 seem unambiguous, but it is important to specify if the sequences are labelled and mutations are ordered. If the sequences are unlabelled, then a permutation of 4 and 5 will create a different data set although the sequences are identical. In Figure 2.10 the mutations are labelled by position, so at level 2 there are three configurations with one sequence identical to the ancestral sequence and one sequence with one mutation on it. If this labelling is removed, then these three configurations are equivalent. This representation is more compact and removes the complication that the data also specifies the positions of the mutations. Any specific position is a point on the real line and the probability that a mutation occurs in that position is zero. The example at the top shows that unlabelling will map two different data sets into one. The lower example shows that if both sequences and mutations are unlabelled then at level 2 in Figure 2.10 three states can be mapped into one state. Removing the information of mutation position allows the data configuration to be represented by a gene tree. The leaves on this tree are again unlabelled and the mutation might just as well have been placed on the right branch. If several mutations are positioned on a branch, no temporal ordering of the mutations is implied.

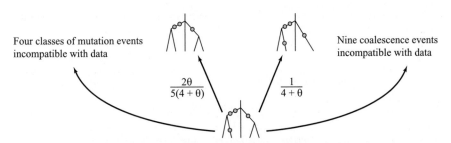

Four classes of mutation events incompatible with data

$$\frac{2\theta}{5(4+\theta)}$$

$$\frac{1}{4+\theta}$$

Nine coalescence events incompatible with data

Figure 2.12 The backward–forward argument is a simple, but clever, way of deriving probability recursions that allows the calculation for at least simple data sets. Go back in time until the first event—a mutation or a coalescence—occurs. The present data configuration could in principle experience $\binom{5}{2}$ coalescence events or a mutation event on one of the five sequences, that is, fifteen possible events. The total rate of all these events would be $\binom{5}{2} + 5\theta/2$ (equation (2.3)). This would allow a probability recursion to be written for the present data set in terms of these fifteen predecessors. However, only two of these fifteen events are compatible with the data—the single mutation on the third sequence can be removed or a coalescence of the two identical sequences 4 and 5; so thirteen of the fifteen possible classes of events can be discarded.

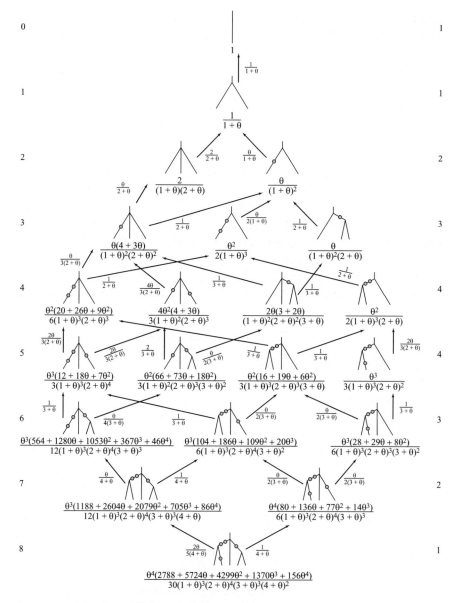

Figure 2.13 This shows all the probabilities and rates necessary to calculate the probability of the present data set with unordered mutations. At the root the ancestral sequence will be assigned probability 1, since it is assumed known. The probability associated with any gene tree is the weighted sum of the probabilities of the gene trees being pointed from that position. The left column of numbers indicates how many events have happened relative to the ancestral sequence. The right column of numbers is the number of sequence configurations for a given number of events in the left column. For rows 2–5 the numbers are smaller than in Figure 2.10, because certain of the configurations have been lumped, that is, mutations have been unordered to ease the presentation (Figure 2.11). The total number of paths through this graph is thirty-three in contrast to Figure 2.10 where it is seventy-two.

does the probabilities that will be derived in Figure 2.13. If the model had been a sequence 1000 bp long with four nucleotides, the full history would have had $4^{1000} \approx 10^{600}$ states and would have been intractable for exact methods. However, it would have been wasteful as a model for human sequences, as most of these states would have vanishing probabilities and so would histories with more than one mutation in a position.

In Figure 2.13 the probabilities and rates of these configurations are all shown. The probability of a configuration can be calculated if the probabilities of the configurations that it could arise from (forward in time) are known by weighing these probabilities with the rates at the corresponding arrow. Which of these paths that are maximally and minimally contributing in probability is dependent on θ. The probability of the present configuration could also be obtained by adding the probability of individual paths, although this is not a computationally rational way of doing this.

The probability of the present configuration is not easy to interpret completely, since it has been reduced, but certain parts are recognisable. In the contribution of each path the denominators will have eight factors of the form $(i - 1 + \theta)$. The power of this factor will count the time spent in a configuration with i sequences (i-sequence configuration). Since mutations do not change the number of sequences, k mutations will imply k times in a configuration with the same number of sequences. In the reduced formula the maximal number of traversals of any path coming from an i-sequence configuration can be seen in $(i - 1 + \theta)^k$, and easily verified on the diagram. For instance four arrows are leaving a 3-sequence configuration. In this case, this means that three of the mutations can occur from a 3-sequence configuration, since the last edge giving a factor $(i - 1 + \theta) = (2 + \theta)$ is an arrow implicating a coalescence.

If the value that maximises the probability of the data ($\theta = 2.12$) is used then the data has probability 0.000744 (see Figure 2.14). If the probabilities of all (infinitely many) data sets with five sequences were calculated, they would sum to one. As the number of sequences is increased each sample configuration will typically have a smaller probability.

2.4.3   Impossible ancestral states

If the ancestral states are unknown, each assignment of ancestral states to segregating sites has equal probability $1/2^k$, if there are k sites. For each assignment, the probability of obtaining the sample configuration can be calculated using the recursion (2.27) and the sum of these weighted by $1/2^k$ can be found. It is the probability of the sample configuration when the ancestral states are unknown. Some assignments are impossible, leading to configurations that are incompatible with the infinite sites assumption that only one mutation is allowed in any site.

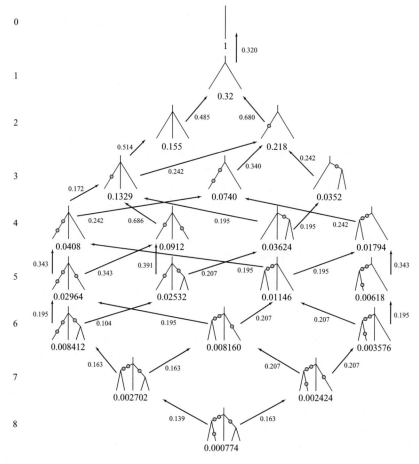

Figure 2.14   This is the same graph as the last figure but this time with $\theta = 2.12$ that makes the data most probable.

There are two levels at which we can address incompatibility with the infinite sites assumption. At a first level we can ask whether there is any assignment of ancestral states to nucleotides that are compatible with a tree. There is a simple test for this, called the four-gamete test: If two columns in the alignment display four different types, 00, 01, 10, and 11, then there cannot be a tree relating the sequences. Note that whatever the assignment of states are, two columns displaying all four types in one assignment show all four types in any assignment. For example, if 0 and 1 are swapped in the first site in 00, 01, 10, and 11, then the types become 10, 11, 00, and 01; again all four types are present.

Given that the sample is compatible with the infinite sites assumption we can ask about how many assignments are compatible. For a given assignment this is easy to test; to calculate the total number of

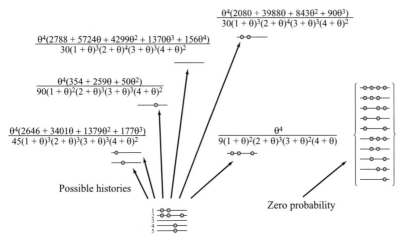

Figure 2.15   If the ancestral allele is not known then the probability of the data can be obtained by summing over all possible ancestral alleles. Since there are four segregating sites, there are sixteen such possibilities that each has probability 1/16. However, only six of these are possible without postulating recombinations in the evolutionary history. The other ten are shown to the right in the figure. If the top sequence (four bullets) was chosen and added to the data, then there would be an incompatibility between positions 2 and 3. Given the data the ten ancestral sequences are impossible and will be assigned probability zero.

compatible assignments is difficult and one cannot derive a simple expression for this.

In the example discussed above, all the balls were chosen as derived mutations, but other choices are conceivable. Since there are four mutations there are $2^4 = 16$ logically possible ancestral states. However, only six of these are possible without recombination as illustrated in Figure 2.15. Some ancestral sequence types are precluded as they would force recombination in the data set. Anticipating Chapter 5 about recombination if a pair of sites exists where all four site configurations are seen, then recombination must have occurred. If the first precluded ancestor sequence in Figure 2.15 is added to the initial configuration, then sites 2 and 3 would have all four possible combinations. The possible ancestral sequences include the four allele types in the present configuration and two configurations, where the first and the second balls are ancestral states.

The probability of the data as a function of ancestral sequences is illustrated in Figure 2.16. If all sixteen possible ancestral states a priori have the same probability of occurring, the probability after having observed the present configuration can be calculated. For this data set, the maximum likelihood estimate of θ depends very little on the chosen ancestral state.

The four-gamete test and compatibility with a tree will be taken up again in Chapter 5.

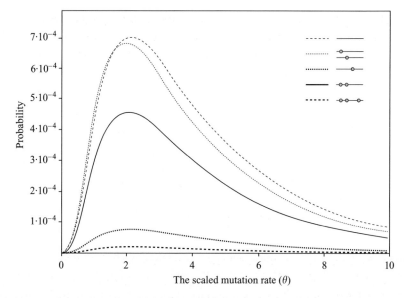

Figure 2.16   The probability of the data set is here seen as a function of θ. Each probability is the sum of the contribution of all paths from the ancestral allele to the present data set. The contribution of each path is determined by the order of the mutations and the coalescent events and the value of this is θ dependent. A more important determinant of the probability is how many paths lead from the ancestral sequence to the present data set. The ancestral configuration 1101 (1 is the mutant state) has only one legitimate path and correspondingly has a very low probability. Figure by Yun Song, personal communication.

2.5   Quantities related to the infinite sites model

2.5.1   The number of segregating sites

The number S_n of mutations in a sample history fulfils a recursion very similar to the recursion for K_n,

$$P(S_n = k) = \frac{\theta}{n - 1 + \theta} P(S_n = k - 1) + \frac{n - 1}{n - 1 + \theta} P(S_{n-1} = k), \quad (2.28)$$

with $k \geq 0$, and initial condition $P(S_1 = 0) = 1$, as there are no segregating sites in one sequence. Alternatively, one could use

$$P(S_2 = k) = \left(\frac{\theta}{\theta + 1} \right)^k \frac{1}{\theta + 1} \qquad (2.29)$$

for $k = 0, 1, \ldots$, as initial condition. It is readily derived from (2.28).

Note the very important difference here when comparing with the recursion for K_n: A sequence experiencing a mutation is not removed but continues to be in the sample; only the number of mutations is counted down by one. Under the infinite alleles model the fate of a sequence does not matter as soon as we know it is a singleton created by mutation. Under the infinite sites model the same sequence might experience several mutations that count in the total number S_n.

The above recursion can also be solved explicitly to yield

$$P(S_n = k) = \frac{n-1}{\theta} \sum_{i=1}^{n-1} (-1)^{i-1} \binom{n-2}{i-1} \left(\frac{\theta}{i+\theta} \right)^{k+1}. \qquad (2.30)$$

In particular the probability of no segregating sites reduces to

$$P(S_n = 0) = \frac{(n-1)!}{(\theta+1)\cdots(\theta+n-1)}. \qquad (2.31)$$

In contrast to K_n there is more information in the sample configuration than can be summarised by S_n. One such example is shown in Figure 2.17, two samples of size four with the same number of mutations, namely one. The relative likelihood curves for the two samples are shown in Figure 2.18.

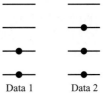

Figure 2.17 Small data examples. The samples have one segregating site, but in the first example the sample is divided into two groups each with two sequences, while in the second example one group is of size three, the other of size one.

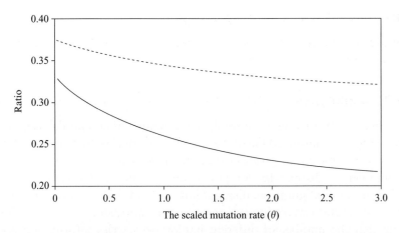

Figure 2.18 Relative likelihood curves. The lower curve shows the relative likelihood curves for the two data samples in Figure 2.17 assuming the ancestral state of the mutations are unknown. The upper curve shows the relative likelihood curve for the second sample assuming the bullet (see Figure 2.17) represents the ancestral state and the mutant state, respectively.

The curve depends on θ implying that the configuration itself carries information about θ that is not captured by S_n.

The variable S_n has mean and variance given by

$$E(S_n) = \theta \sum_{j=1}^{n-1} \frac{1}{j},$$ (2.32)

and

$$\text{Var}(S_n) = \theta \sum_{j=1}^{n-1} \frac{1}{j} + \theta^2 \sum_{j=1}^{n-1} \frac{1}{j^2}.$$ (2.33)

Let η_i be the number of mutation in frequency i, $i = 1, \ldots, n-1$. Then $\sum_{i=1}^{n-1} \eta_i = S_n$, and the mean of η_i is

$$E(\eta_i) = \frac{\theta}{i}.$$ (2.34)

If the ancestral state of a segregating site is not known then mutations found in frequency i and $n - i$ cannot be distinguished. In particular, the number of singletons (one of the two alleles is in frequency one) has expectation

$$\theta \left(1 + \frac{1}{n-1} \right).$$ (2.35)

All other sites are called informative sites, i.e. sites where both alleles are found in at least two copies in the sample.

2.5.2 Haplotypes

It is often more convenient to think of the infinite alleles model as a model of haplotype evolution. Under the assumption that mutations never happen twice in the same position (infinite sites assumption), K_n is exactly the number of distinct haplotypes in the sample, and $(a_1, a_2, \ldots, a_n)$ is the haplotype configuration, disregarding any additional information about how many mutations distinguish one type from another type. It is thus interesting that the number of different haplotypes carries information about θ, whereas the particular allele configuration given K_n is uninformative about θ (equation (2.23)).

The number of mutations that increases the number of haplotypes is $K_n - 1$. In consequence those mutations that do not increase K_n amount to

$S_n - (K_n - 1)$, which has mean

$$E(S_n) - E(K_n) + 1 = \theta \sum_{j=1}^{n-1} \frac{1}{j} - \sum_{j=0}^{n-1} \frac{\theta}{j+\theta} + 1 = \sum_{j=1}^{n-1} \frac{\theta^2}{j(j+\theta)}. \quad (2.36)$$

This difference is bounded (less than $2\theta^2$) even for large sample sizes, despite the fact that $E(S_n)$ and $E(K_n)$ grow towards infinity for increasing n. Thus, in a large sample most mutations create novel haplotypes.

Singleton mutations can be one of two kinds. First, a mutation can be the only mutation unique to a haplotype. Second, it can be one of several mutations unique to a particular haplotype. The mean number of mutations of the second kind can be found using equations (2.26) and (2.35),

$$E(\eta_i) - E(a_1) = \frac{n\theta^2}{(n-1)(n-1+\theta)}. \quad (2.37)$$

For example, if $n = 20$ and $\theta = 10$ then there are on average 3.63 singleton mutations hitting branches with already one singleton mutation. In total there are 10.53 singleton mutations and 6.90 haplotypes that occur in one copy only.

2.5.3   Pairwise mismatch distribution

The distribution of the number of segregating sites (also called mismatches) between two sequences is given in equation (2.29). It is a modified geometric distribution (it can take the value 0) with parameter $1/(1+\theta)$. If all possible pairs in a sample are compared then the distribution of the pairwise differences, $\binom{n}{2}$ pairs in total, does not look like a geometric for two reasons: The first is that all n sequences share genealogical history and are thus not independent. The second is that each sequence is used in $n-1$ comparisons. A few examples are given in Figure 4.5. The distribution is typically bimodal because of the basal split in the genealogy at the root. If there are k and $n-k$ sequences, respectively, hanging from the root then the $\binom{k}{2}$, respectively $\binom{n-k}{2}$, comparisons in the first, respectively second, group tend to be smaller than the $k(n-k)$ intergroup comparisons. This distinctive feature disappears in some coalescent models that are introduced in Chapter 4. The mismatch distribution might thus be used as a test of the appropriateness of the basic coalescent model to describe the data sample.

2.5.4   Estimators of θ and Tajima's *D*

Let π_{ij} be the number of differences between two sequences, i and j, and

$$\hat{\pi} = \frac{2}{n(n-1)} \sum_{i,j} \pi_{ij}, \tag{2.38}$$

the average over all pairs in a sample. $\hat{\pi}$ is an estimator of θ with mean

$$E(\hat{\pi}) = \theta, \tag{2.39}$$

because π_{ij} has mean θ. This estimator is known as Tajima's estimator of θ. Another estimator of θ is Watterson's estimator defined by

$$\hat{\theta}_W = \frac{S_n}{a_n}, \tag{2.40}$$

where

$$a_n = \sum_{i=1}^{n-1} \frac{1}{i}. \tag{2.41}$$

Also $\hat{\theta}_W$ has mean θ (equation (2.32)). Both estimators are further discussed in Section 6.2. If $\hat{\pi}$ and $\hat{\theta}_W$ are compared their difference is expected to be close to zero and deviations from zero are thus indicative of model misspecification, that is, the basic coalescent model does not capture the variation in the data. Tajima (1989) normalised the difference in what is now known as Tajima's *D*

$$D = \frac{\hat{\pi} - \hat{\theta}_W}{\sqrt{e_1 S_n + e_2 S_n (S_n - 1)}}. \tag{2.42}$$

Here

$$e_1 = \frac{n+1}{3a_n(n-1)} - \frac{1}{a_n^2}, \tag{2.43}$$

$$e_2 = \frac{1}{a_n^2 + b_n} \left(\frac{2(n^2 + n + 3)}{9n(n-1)} - \frac{n+2}{na_n} + \frac{b_n}{a_n^2} \right), \tag{2.44}$$

and

$$b_n = \sum_{i=1}^{n-1} \frac{1}{i^2}. \tag{2.45}$$

This normalisation of the difference ensures that *D* has mean close to zero and variance close to one, but not that it follows a normal distribution.

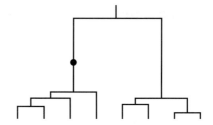

Figure 2.19   In the first example each mutation is counted in $n - 1$ comparisons. In the second each mutation is counted in $(n/2)(n/2) = n^2/4$ comparisons, assuming the subtrees have equal size. Here $n = 8$.

In fact, Tajima showed that the distribution is close to a beta distribution. If D calculated on empirical data fulfils $|D| \leq 2$ it is usually taken to imply that the basic coalescent explains the data well.

Two extreme violations of the basic coalescent can be imagined. In the first case all sequences sprung from a common ancestor and evolved thereafter independently of each other (Figure 2.19). In that case all mutations are counted $n - 1$ times in $\hat{\pi}$ and consequently

$$\hat{\pi} - \hat{\theta}_W = \frac{2}{n} S_n - \frac{S_n}{a_n} < 0 \tag{2.46}$$

for $n > 2$. In the second case all but the last coalescent event happened near the present time (Figure 2.19). If the number of genes subtending the two lineages at the basal split is $n/2$ then

$$\hat{\pi} - \hat{\theta}_W = \frac{n}{2(n - 1)} S_n - \frac{S_n}{a_n} > 0, \tag{2.47}$$

assuming all mutations happen on the two long branches. (See also Chapter 6.)

2.6   Evolutionary versus sampling variance

In the basic coalescent the common ancestry is modelled as a stochastic element. Therefore, the variance of a variable, for instance S_n, is attributable to two sources: (1) random sampling of mutations from n genes with a given, but unknown, ancestral relationship; and (2) the stochastic nature of the ancestry of the n genes. This is illustrated in Figure 2.20. Also, the two sources are transparent in the way sequence data typically are simulated: In the first step the genealogy itself is simulated; in the second mutations are superimposed onto the genealogy.

The variance of a variable, in the example S_n, can be decomposed into two components that reflect the two sources of variation. The first component

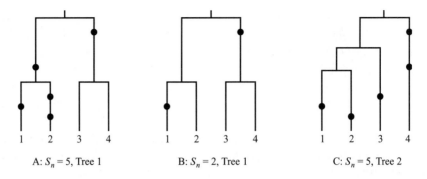

A: $S_n = 5$, Tree 1   B: $S_n = 2$, Tree 1   C: $S_n = 5$, Tree 2

Figure 2.20   Two sources of variation: Sampling of mutations along the genealogy of the sample, and sampling of the genealogy itself.

is called the *sampling variance* and the second component the *evolutionary variance*. By sampling variance the following is understood. Let X be the variable of interest. Consider the variance, $\text{Var}(X \mid \mathcal{G})$, of X for a given (fixed) genealogy $\mathcal{G}$. For example, if $\mathcal{G}$ is tree A in Figure 2.20 and X is S_n, then $\text{Var}(X \mid \mathcal{G}) = \text{Var}(S_n \mid \text{Tree A})$ is the variance in the number of mutations given that the genealogy relating the four sequences is the tree A. That is, $\text{Var}(S_n \mid \text{Tree A})$ is the variance due to sampling of mutations only.

Now, $\text{Var}(X \mid \mathcal{G})$ depends on the genealogy $\mathcal{G}$. To obtain the sampling variance, $V_S(X)$, of X, the average of $\text{Var}(X \mid \mathcal{G})$ over all possible genealogies, $\mathcal{G}$, is computed, that is,

$$V_S(X) = E[\,\text{Var}(X \mid \mathcal{G})\,]. \tag{2.48}$$

If the variance of X is $\text{Var}(X)$, then the evolutionary variance, $V_E(X)$, is simply defined as

$$V_E(X) = \text{Var}(X) - V_S(X), \tag{2.49}$$

such that $V_S(X)$ and $V_E(X)$ together add up to $\text{Var}(X)$. These two definitions are exemplified below.

2.6.1   Example 1: The variable S_n

The variance of S_n is

$$\text{Var}(S_n) = \theta \sum_{j=1}^{n-1} \frac{1}{j} + \theta^2 \sum_{j=1}^{n-1} \frac{1}{j^2}. \tag{2.50}$$

For a given tree ($\mathcal{G}$) the distribution of S_n depends only on the total branch length, L_n, of the tree. In fact, S_n is Poisson distributed $\text{Po}(\theta L_n/2)$, and hence the variance of S_n *given a particular tree* is $\theta L_n/2$. Averaged over all possible trees this amounts to $\theta \sum_{j=1}^{n-1} 1/j$, because $E(L_n) = 2 \sum_{j=1}^{n-1} 1/j$. Thus, the first term in equation (2.50) is simply the sampling variance of S_n. In consequence, the second term must be the evolutionary variance. In this particular example, the evolutionary variance is also the variance of $\theta L_n/2$.

Which of the two components contributes most to the total variance in (2.50) depends on θ and n. For small n or small θ, an explicit calculation of $V_S(S_n)$ and $V_E(S_n)$ is required, because a change in one of the two parameters might also change the relationship between the two components. For large n, the situation is different. Note that

$$\sum_{j=1}^{n-1} \frac{1}{j^2} \approx \frac{\pi}{6}, \quad \text{and} \quad \sum_{j=1}^{n-1} \frac{1}{j} \approx \log(n), \tag{2.51}$$

and $V_S(S_n)$ will eventually dominate $\mathrm{Var}(S_n)$; that is, $V_E(S_n)$ will be comparably insignificant. For large θ the situation is reversed. Here $V_E(S_n)$ eventually dominates.

Compare this 'two component variance' to the situation in phylogenetic analysis where the tree is an unknown fixed parameter. All variation in the number of mutations comes from sampling mutations along the fixed, but unknown, tree. Going back to Figure 2.20 for a moment, A and B represent two outcomes of S_n with the same genealogical (or phylogenetic) relationship—A and B only differ in the number of mutations added to the tree. If the mutation rate were to be estimated from, say $S_n = 5$, using phylogenetic methods, only the sampling variance would be accounted for, not the variance caused by the stochastic nature of the tree.

2.6.2 Example 2: Tajima's estimator $\hat{\pi}$

Tajima's estimator, $\hat{\pi}$, has variance

$$\mathrm{Var}(\hat{\pi}) = \frac{(n+1)\theta}{3(n-1)} + \frac{2(n^2+n+3)\theta^2}{9n(n-1)}. \tag{2.52}$$

One can show that the sampling variance, $V_S(\hat{\pi})$, is exactly the first term in equation (2.52), and that the evolutionary variance is the second term. For large θ, the $V_E(\hat{\pi})$ is dominating, just as in the previous example. For large n the situation is different. Note that

$$V_S(\hat{\pi}) \approx \frac{\theta}{3}, \quad \text{and} \quad V_E(\hat{\pi}) \approx \frac{2\theta^2}{9}. \tag{2.53}$$

Which one is the larger depends on θ more than n: If $\theta < 3/2$, then $V_S(\hat{\pi}) > V_E(\hat{\pi})$, and if $\theta > 3/2$, then $V_S(\hat{\pi}) < V_E(\hat{\pi})$.

Recommended reading

Ewens, W. J. (2004), *Mathematical Population Genetics*, 2nd edn, Springer Verlag.

Jukes, T. H. and Cantor, C. R. (1969), Evolution of protein molecules, *in* H. N. Munroe, ed., 'Mammalian Protein Metabolism', Academic Press, pp. 21–132.

Griffiths, R. C. and Tavaré, S. (1995), 'Unrooted genealogical tree probabilities in the infinitely-many-sites model', *Math. Biosci.* **127**(1), 77–98.

Tavaré, S. (1984), 'Line-of-descent and genealogical processes, and their applications in population genetics models', *Theor. Popul. Biol.* **26**(2), 119–164.

3 Trees and topologies

3.1 Some terminology

The simplicity of the coalescent process allows various quantities to be calculated from combinatorial arguments. This is the focus of the present chapter. However, first we will specify some concepts regarding coalescent trees and their relationships to the well-known concept of phylogenetic trees.

3.1.1 The jump process and the waiting time process

In Kingman's original formulation of the coalescent he separated the process into two independent processes, the jump process and the waiting time process. Figure 3.1 illustrates these processes for a sample of five genes. The waiting time process describes the waiting times to the next coalescent event when there are k genes left and is described by exponentially distributed variables as derived in Chapter 1. The jump process describes which pair of genes coalesce at each coalescence event and was originally phrased in terms of equivalence classes. An equivalence class for a set of genes is a specific separation of the set of genes into disjoint sets covering the whole set of genes. In Figure 3.1 the first equivalence class is the separation of the genes into five different sets. After the first coalescence event, two genes (4 and 5) are merged into the same set. The jump process thus determines a path through equivalence classes from the initial state to the final state which consists of a single set containing all the genes. The path through equivalence classes uniquely defines the coalescence events that occurred in a time ordered fashion. This is termed the coalescent topology.

3.1.2 The coalescent and phylogenetic trees

For any set of homologous genes not subject to recombination, there exists a phylogeny with dates at inner nodes, corresponding to the splitting events (formulated in the forward view—viewed backward in time, it would be coalescent events) that created the present genes. There would also be a root with a date corresponding to the MRCA of the genes. This phylogeny

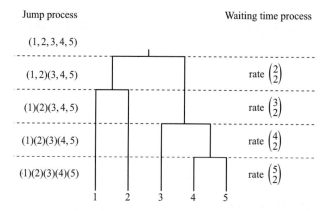

Figure 3.1   The coalescent tree. The coalescent process generating coalescent trees can be divided into two independent processes: one process describing the waiting times between events, another describing which genes coalesce. The latter is known as the jump process.

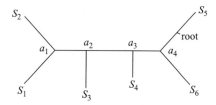

Figure 3.2   Basic terminology. A tree is represented by a set of k nodes and $k - 1$ branches or edges. Nodes with only one edge attached are called leaves, tips, or external nodes; all other nodes are called internal nodes and have three edges attached. In the example there are $k = 10$ nodes and $k - 1 = 9$ edges. Of the ten nodes, six are external $(S_1, \ldots, S_6)$ and four are internal $(a_1, \ldots, a_4)$. The root breaks an edge into two edges. In the example, adding the root increases the number of edges by one. Labels at internal nodes can be permuted without changing the biological interpretation of the tree.

exists, but it is likely not to be obtainable from the present genes, because of lack of information in the genes. Real nucleotide data have evolved on this phylogeny according to a mutation process (see Chapter 2). Some basic terminology is shown in Figure 3.2.

3.1.2.1   *Phylogenetic trees*

In the field of interspecific phylogenetics, an example tree may look like Figure 3.3. The sampled DNA sequences are analysed under a DNA substitution model (e.g. Jukes–Cantor model, see Chapter 2) and a tree is reconstructed using a reconstruction principle, for example, parsimony or maximum likelihood. The root is usually determined using an outgroup species. The resulting tree will have its branches measured in units of substitutions. The root defines the direction of time, but there is no information on the relative times of the other two internal nodes. This is only possible if a

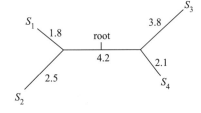

Figure 3.3   A rooted phylogenetic tree of four genes. The branch lengths are in units of substitutions. The root determines the direction of time in the phylogeny. In this case, a molecular clock is not assumed.

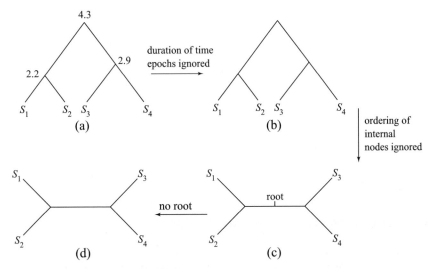

Figure 3.4   Four different representations of a coalescent tree with different levels of information. (a) A fully determined coalescent tree with root and dates on inner nodes. (b) A coalescent tree with no information on the dates of inner nodes but with the correct ordering of coalescent events, the coalescent topology. (c) The tree without information on the ordering of coalescent events, a rooted tree topology. (d) Tree without a root, an unrooted tree topology.

molecular clock can be imposed on the tree, in which case it will resemble the coalescent tree.

3.1.2.2   *Coalescent trees*

The contrast between the phylogenetic tree and a coalescent tree is illustrated by Figure 3.4. In a coalescent tree branch lengths are always proportional to calender time (calender time = coalescent time times $2N$ generations times generation time) and all contemporary genes have the same amount of time to the MRCA.

The mutation models presented in Chapter 2 obey the molecular clock. Mutations happen at rate $\theta/2$, irrespective of whether an infinite alleles, infinite sites, or finite sites model is assumed. In all cases the relationship between coalescent time and expected number of mutations is

$$\text{Expected number of mutations} = \frac{\theta}{2}\text{Coalescent time} \qquad (3.1)$$

(equation (2.6)). To show a coalescent tree in expected number of substitutions we just have to multiply all branch lengths by $\theta/2$.

As one moves from the coalescent tree (a) in Figure 3.4 to tree (d) fewer and fewer constraints are imposed on the tree. In (a) all internal nodes are ranked in time: The event merging the ancestor of S_1 with that of S_2 is closer towards the present time than the event merging the ancestors of S_3 and S_4. In (b) the coalescent topology is retained but the actual time of coalescence events (the waiting time process) is not included. In (c), the relative timing of the coalescence events is ignored, and a rooted tree topology similar to the phylogenetic tree of Figure 3.3 results. In (d), neither root nor time is given. If the times of the two coalescent events in (a) are interchanged the unrooted tree topology would still be as in (d). Therefore, different coalescent trees might have the same unrooted topology.

3.2 Counting trees and topologies

The number of coalescent topologies and binary unrooted tree topologies with k leaves are abbreviated, C_k and B_k, respectively (C for coalescent, B for binary). Both obey simple recursions and initial conditions

$$C_k = \binom{k}{2}C_{k-1}, \quad C_2 = 1, \qquad (3.2)$$

and

$$B_k = (2k - 5)B_{k-1}, \quad B_3 = B_2 = 1. \qquad (3.3)$$

The recursion for C_k is almost unnecessary since it follows directly from the $k - 1$ consecutive choices that have to be made to define the coalescent topology. The recursion for B_k is illustrated in Figure 3.5. There is a one-to-one correspondence between an unrooted tree topology and a list of choices of adding the edges leading to leaves, $1, 2, \ldots, k$. The kth leaf must be placed on one of the $2k - 5$ edges on an unrooted tree topology of a tree relating $k - 1$ leaves. This tree will then gain one leaf and two edges. The recursions can be solved to give:

$$C_k = \prod_{j=2}^{k} \frac{j(j - 1)}{2} = \frac{k!(k - 1)!}{2^{k-1}}, \qquad (3.4)$$

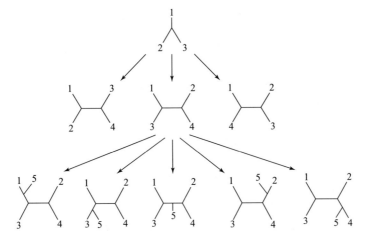

Figure 3.5 The basic recursion for the number of unrooted tree topologies as a function of leaves. Here, k goes from three to five. $k = 3$: One topology. $k = 4$: A topology with three $(k - 1 = 3)$ leaves has $2k - 5 = 2 \cdot 4 - 5 = 3$ branches implying that there are three possible topologies for four leaves. $k = 5$: An unrooted tree with four leaves $(k - 1 = 4)$ has $3k - 5 = 10 - 5 = 5$ branches, so the fifth leaf can be added in five ways yielding in total of $3 \cdot 5 = 15$ possible topologies.

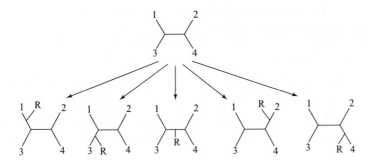

Figure 3.6 The number of rooted trees for an unrooted topology with four leaves is five since the root (R) can be placed on each of the five internal branches.

and

$$B_k = \prod_{j=3}^{k} (2j - 5) = \frac{(2k - 5)!}{2^{k-1}(k - 3)!} = (2k - 5)!!. \qquad (3.5)$$

The quantity $(2k - 5)!!$ is not the factorial of $(2k - 5)!$ but a way to describe a factorial-like number that only multiplies every second integer, for example, $5!! = 5 \cdot 3 \cdot 1$. The number, $\mathcal{R}_k$, of rooted topologies equals B_{k+1} because any unrooted topology of k leaves can be made into a rooted topology by choosing one of the possible $2k - 5$ branches to place the root, see Figure 3.6.

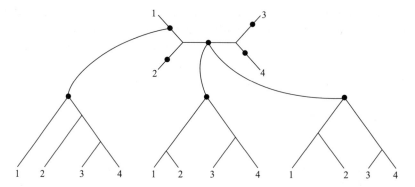

Figure 3.7   From unrooted trees to coalescent trees. There are five possible edges to place the root. If the root is placed on an external edge it gives rise to one coalescent tree. If the root is placed on the internal edge it gives rise to two coalescent trees—one in which 1 and 2 merge before 3 and 4, and one in which 3 and 4 merge before 1 and 2.

Table 3.1   The number of topologies[a]

k	2	3	4	5	6	8	10	15	20
$\mathcal{B}_k$	1	1	3	15	105	$1.0 \cdot 10^4$	$2.0 \cdot 10^6$	$7.9 \cdot 10^{12}$	$2.2 \cdot 10^{20}$
$\mathcal{C}_k$	1	3	18	180	2700	$1.6 \cdot 10^6$	$2.6 \cdot 10^9$	$7.0 \cdot 10^{18}$	$5.6 \cdot 10^{29}$

[a] The table shows the number $\mathcal{C}_k$ of coalescent topologies, and the number $\mathcal{B}_k$ of unrooted topologies. The number $\mathcal{R}_k$ of rooted topologies for k leaves equals $\mathcal{B}_{k+1}$.

Each unrooted tree topology corresponds to a number of (rooted) coalescent trees. How many coalescent trees each unrooted topology covers depends on the chosen root edge and the shape of the tree. As an example consider the topology for four leaves in Figure 3.7. If the root is placed on any of the four external edges there is only one corresponding coalescent topology; if the root is placed in the only internal edge there are two possible coalescent trees.

The number of topologies as a function of the number of leaves is shown in Table 3.1. It is clear that the number of topologies relating a set of genes grows very fast indeed. This is a genuine problem, since most methods of evaluating the probability of data should in principle investigate these topologies one by one. There is no simple way of handling them all at once, in the same way that it is possible to integrate some functions over a real interval instead of summing over all reals in the interval.

3.3   Gene trees

A gene tree is a visual representation of the data. More specifically, it is a graph that shows the ancestral relationships between genes. In Chapters 1 and 2 we used gene trees to illustrate data sets and recursions to calculate the probability of a data set. To build a gene tree the infinite sites model is assumed. In Chapter 5 we discuss what happens when this is not the case. Each mutation in the sample imposes a partition of the sequences into two sets according to the genetic type they carry. The gene tree can be built from this information alone, either in a manual way starting out with the first segregating site in the sample and then moving on to the next site, and so forth; or by using an algorithm originally developed by Gusfield (1991). The algorithm is stated below.

Table 3.2 gives an example of a small data set. The first site imposes the partition of genes into the two sets {1, 2, 4} and {3, 5}. This is illustrated in Figures 3.8 and 3.9.

Gene trees are convenient visualisations of data. Some remarks are needed. First of all, the gene tree is not a coalescent tree, but a graph that clusters genes according to their type and mutation pattern. Therefore, the gene tree is not necessarily bifurcating, as seen in Figure 3.9. Also the gene tree does not tell whether sequences 2 and 4 merge before merging with the ancestor of sequence 1, or whether sequences 1, 3, and 5 find a MRCA before coalescing with 2 and 4. However, all coalescent trees consistent with the sample must agree with the clusters in the gene tree.

Table 3.2   A data set with five sequences and four segregating sites with relative positions[a]

	0.012	0.41	0.47	0.97
1	0	0	0	0
2	0	0	1	1
3	1	0	1	0
4	0	0	1	1
5	1	1	1	0

[a] The sites partition the sequences into the following sets: {1, 2, 4}, {3, 5}; {1, 2, 3, 4}, {5}; {1}, {2, 3, 4, 5}; and {1, 3, 5}, {2, 4}.

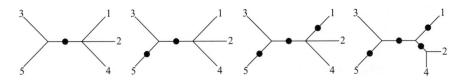

Figure 3.8   A gene tree representing the data set in Table 3.2. The figure illustrates how the gene tree is built up, starting with the first site, gradually adding more and more sites to the tree. A gene tree can be represented in many ways—here sequences are represented with a branch leading to the sequence even if no mutations occur on the branch. Filled circles denote mutations.

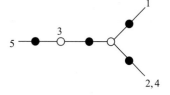

Figure 3.9   The gene tree in Figure 3.8 with branches without mutations removed. This caused sequence 3 to be placed on the branch connecting sequence 5 with the rest of the sample. Filled circles denote mutations and open circles (ancestral) sequences.

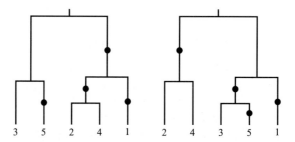

Figure 3.10   The gene tree in Figure 3.8 rooted in two different places, resulting in two different trees. These trees can be multifurcating and the order of events is not fixed.

Second, spatial information is lost in the gene tree, unless all mutations are labelled with their position in the sequences. If several mutations occur on a single branch then their order is arbitrary; whether one occurred before the other cannot be judged from the information in the gene tree.

Third, the data can be consistent with a gene tree even though the sequences that have been subject to recombination or multiple mutations have happened in the same position. The effect of recombination on gene trees will be discussed in Chapter 5. In these cases the gene tree cannot be taken to represent evolution as it actually occurred. But what is the gene tree then? The gene tree can be said to be the most parsimonious history of the data in terms of the number of mutation and recombination events, because it shows evolution as if there were no recombination and all mutation events were unique.

Finally, the only assumption that is imposed (apart from no recombination) is the infinite sites assumption: at most one mutation in each column of the alignment. Thus, it does not matter whether the sequences have evolved under selection or whether the size of the population has fluctuated over time, been through bottlenecks or something else. This makes the gene tree an extremely convenient and useful concept.

Knowledge about the state of ancestral nucleotides can be used to root the tree. This information might be obtained from a closely related species, for example, a gene tree of human sequences might be rooted using a chimpanzee sequence. Return for a moment to Figure 3.8: The interpretation of the tree depends on where the root is placed, two examples are given in Figure 3.10. In the first rooted tree sequences 1, 2, and 4 carry a common mutant type, in the second rooted tree sequences 3 and 5 carry a common mutant type.

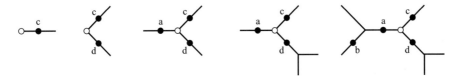

Figure 3.11   The construction of a gene tree from the data set in Table 3.2 using Gusfield's algorithm.

3.3.1   How to build a gene tree

Gusfield (1991) provided an elegant algorithm to build a gene tree. He showed that there is a gene tree if and only if there is a gene tree with the most frequent type in each column as the ancestral type. Thus, the first step is to recode the sequences such that the most frequent type in each column is zero. If there is ambiguity, that is, both 0 and 1 occur in the same frequency, choose one of them arbitrarily. Here, the algorithm is illustrated on the data underlying Table 3.2, and the stepwise building of the gene tree is illustrated in Figure 3.11.

	a	b	c	d
1.	0	0	1	0
2.	0	0	0	1
3.	1	0	0	0
4.	0	0	0	1
5.	1	1	0	0.

Identical columns are left out (there are none in this example) and sites are labelled with letters. Do the following:

1. Determine whether the data is compatible with a gene tree (see also Chapter 5 for more about compatibility and its relation to recombinations). That is, determine whether all four gametes 00, 01, 10, and 11 are present in any two columns. If they are there cannot be a gene tree. Otherwise continue.

2. Considering each column as a binary number, sort the numbers into decreasing order, with the largest number in column one. Continuing with the example:

	c	d	a	b
1.	1	0	0	0
2.	0	1	0	0
3.	0	0	1	0
4.	0	1	0	0
5.	0	0	1	1.

3. The gene tree is constructed by adding a sequence with all its characters one at a time. The characters of a sequence to be added is a specific row, which is read from right to left. The sequence is placed by tracing from the leaves towards the root. It has its own edge until the prefix is encountered where it coincides with the last added character.

The root is labelled with an open circle. It can be removed to form an unrooted tree.

3.4   Nested subsamples

An interesting variant of the coalescent process was studied by Saunders et al. in 1984. Assume a sample, A, is taken of size n, and within that sample a subsample, B, of size m is also taken. How does the relationship of these two evolve over time? Obviously, either of the samples taken in isolation behaves like the standard coalescent, so the interesting question is their relationship. We have touched on this relationship previously in the description of how a sample can be considered part of still larger and larger samples (Chapter 1).

In the coalescent the process describing the number of ancestors is very simple, it starts in n and then descends until it reaches one. In the nested sampling it is slightly more complicated: the process describing the number of ancestors starts out in (m, n) and then jumps to either $(m, n - 1)$ or $(m - 1, n - 1)$. The first situation occurs if the coalescent event involves at least one gene that is not in B—it could be one or two (see Figure 3.12). The second case occurs if both genes are from B. The scheme can be generalized to a hierarchy of samples within samples, but this will not be pursued here.

The jump process can be described analytically in the following way. Assume the process is in state (i, j). Then

$$(i, j) \rightarrow (i - 1, j - 1) \text{ with probability } \frac{i(i - 1)}{j(j - 1)}, \qquad (3.6)$$

and

$$(i, j) \rightarrow (i, j - 1) \text{ with probability } 1 - \frac{i(i - 1)}{j(j - 1)}. \qquad (3.7)$$

Note that if $i = 1$ then every jump is of the form $(i, j) = (1, j) \rightarrow (1, j - 1)$. A specific outcome of the process has probability given by multiplication of individual terms, for example, $(3, 6) \rightarrow (2, 5) \rightarrow (1, 4) \ldots$ has probability $(3 \cdot 2)/(6 \cdot 5) \cdot (2 \cdot 1)/(5 \cdot 4) \cdots = 0.02 \cdots$.

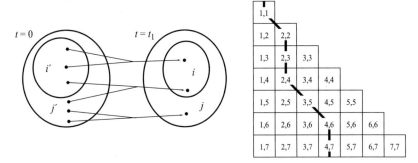

Figure 3.12 The coalescent process for nested subsamples. Consider a sample, A, of size n and a subsample, B, of size $m \leq n$, embedded in A. At each coalescent event there are three possibilities: (1) two genes in B find a common ancestor; (2) one gene in B and one not in B find a common ancestor, and (3) two genes not in B find a common ancestor. In this case the ancestor will not be an ancestor of B. This process starts in the integer pair (m, n) and jumps down through possible pairs (i, j) until it reaches $(1,1)$ when all genes have found a common ancestor. In the example $(m, n) = (4, 7)$ and the process jumps to $(4, 6)$, then to $(3, 5)$—the first event is of type (2) or (3), either is possible, the next is of type (1).

A series of interesting questions can be answered in this framework. For example, what is the probability that the MRCA of B is also the MRCA of A? The answer is

$$\frac{(n+1)(m-1)}{(n-1)(m+1)}. \tag{3.8}$$

Some special cases are of interest. If A is the whole population ($n \approx \infty$, or $n = 2N$ and $2N$ is large), the probability in equation (3.8) becomes

$$\frac{m-1}{m+1}, \tag{3.9}$$

which is high even for moderate sample sizes of B. Thus, the MRCA of B is the MRCA of the whole population with high probability. For a sample size of nine the probability is 0.8. If A is not the whole population then this probability is higher than 0.8. If $n = 19$ then the probability is 0.89. The intuitive reason behind equation (3.8) is the quadratic increase in coalescence rate with sample size, which makes the large sample shrink much faster than the small sample. Within a short time the probability that their ancestors are the same becomes quite high.

If $m = 2$ then the MRCA of B is the whole populations MRCA with probability $1/3$. Remember the time until the whole population has found a MRCA is two (in coalescent units), and the time until a sample of size two has found a MRCA is one. Thus,

$$\frac{1}{3}E[T_2 \mid A = B] + \frac{2}{3}E[T_2 \mid A \neq B] = 1, \qquad (3.10)$$

where $A = B$ (and $A \neq B$) is short for 'the MRCA of A is that of B' (and 'the MRCA of A is not that of B', respectively), or

$$E[T_2 \mid A \neq B] = \frac{3}{2}\left(1 - \frac{1}{3}E[T_2 \mid A = B]\right) = \frac{3}{2}\left(1 - \frac{2}{3}\right) = \frac{1}{2}, \quad (3.11)$$

because $E[T_2 \mid A = B]$ is the mean time until a MRCA of the whole population. Thus knowing that B does not share a MRCA with the whole population decreases the time until a MRCA of B by a factor of two. Similarly, one can show that the variance is

$$\mathrm{Var}(T_2 \mid A \neq B) \approx 0.17,$$

which is a reduction by a factor of six compared to the (unconditional) variance $\mathrm{Var}(T_2) = 1$.

3.5 Hanging subtrees

In the previous section we took a bottom-up look at the jump process. We started at the present time and followed a subsample back until a MRCA had been reached. In this section we take a top-down look and follow the number of descendants of a lineage. Let a sample of size n be given and consider the time when there were k ancestral lineages. In the example shown in Figure 3.13 k is chosen to be 3 and n chosen to be 8. Just before the coalescent event that reduced the number of lineages to k there were $k + 1$ lineages.

When looked at from a top-down perspective all k lineages have equal chance of splitting up into two lineages. Also the $k + 1$ lineages have equal chance of splitting up again and so on, until there are n lineages. Assume the n genes are unlabelled but that it is registered how many genes each of the k lineages subtend (i.e. the lineages are ordered). The probability that the k lineages have $n_1, n_2, \ldots, n_k$ descendants, $\sum_i n_i = n$, is

$$P_n(n_1, \ldots, n_k, \text{ unlabelled, ordered}) = \frac{(n-k)!(k-1)!}{(n-1)!}, \qquad (3.12)$$

which does not depend on $n_1, \ldots, n_k$. As a consequence a permutation of an ordered configuration has the same probability as the unpermuted

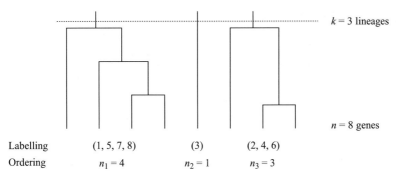

Labelling	$(1, 5, 7, 8)$	(3)	$(2, 4, 6)$
Ordering	$n_1 = 4$	$n_2 = 1$	$n_3 = 3$

Figure 3.13 Hanging subtrees. Shown is the lower part of a genealogy of a sample of size 8 subtending $k = 3$ lineages with $n_1 = 4$, $n_2 = 1$, and $n_3 = 3$, respectively. Below the genealogy is shown one possible labelling of the genes. If labels are ignored the ordering of the subtree sizes remain.

configuration, for example, $(n_1, n_2, n_3) = (3, 3, 2)$ has the same probability as $(n_1, n_2, n_3) = (3, 2, 3)$. Equation (3.12) can be obtained using a forward–backward recursion argument similar to those encountered in Chapter 2, but in this case only coalescence events are relevant.

The configuration of a sample subtending k lineages might be described in one of four ways according to whether genes are labelled/unlabelled and whether the k lineages are ordered/unordered. We state the probabilities for the different combinations below (and give examples in Table 3.3)—the two most important combinations are unordered configurations, unlabelled or labelled. We would rarely be interested in ordered configurations because this level of information would hardly ever be available. However, equation (3.12) is particularly simple to derive, for example, using a recursion approach, and thus a natural starting point.

To obtain the probabilities for labelled configurations note that an ordered configuration $(n_1, n_2, \ldots, n_k)$ can be labelled in

$$\frac{n!}{n_1! n_2! \ldots n_k!} \tag{3.13}$$

different ways. One specific labelling of a sample of size 8 is shown in Figure 3.13. The probability of a particular labelling when the k lineages have $n_1, n_2, \ldots, n_k$ descendants, $\sum_i n_i = n$, is

$$P_n(n_1, \ldots, n_k, \text{labelled, ordered}) = \frac{(n-k)!(k-1)!}{n!(n-1)!} n_1! \cdots n_k!, \tag{3.14}$$

which is equation (3.12) divided by equation (3.13). In the example in Figure 3.13, $n_1 = 4$, $n_2 = 1$, and $n_3 = 3$, and by insertion into (3.14) this yields a probability of $1.7 \cdot 10^{-4}$.

Table 3.3 Subtending configurations of three lineages in a sample of size 8^a

n_1	n_2	n_3	No. of lab	No. of ord	lab, ord	unlab, ord	lab, unord	unlab, unord
6	1	1	56	3	$8.5 \cdot 10^{-4}$	0.048	$5.1 \cdot 10^{-3}$	0.143
5	2	1	168	6	$2.8 \cdot 10^{-4}$	0.048	$1.7 \cdot 10^{-3}$	0.286
4	3	1	280	6	$1.7 \cdot 10^{-4}$	0.048	$1.0 \cdot 10^{-3}$	0.286
4	2	2	420	3	$1.1 \cdot 10^{-4}$	0.048	$6.9 \cdot 10^{-4}$	0.143
3	3	2	560	3	$8.5 \cdot 10^{-5}$	0.048	$5.1 \cdot 10^{-4}$	0.143

^{a} Each of the five unordered, unlabelled configurations n_1, n_2, and n_3 are listed with the number of labellings and orderings, and the probability of a specific ordered, labelled configuration (equation (3.14)), ordered, unlabelled configuration (equation (3.12)), unordered, labelled configuration (equation (3.18)), and unordered, unlabelled (equation (3.17)) configuration. In the first row, for example, the probability 0.048 is the product of 56 and $8.5 \cdot 10^{-4}$, and the probability 0.143 is the product of 3 and 0.048. 'lab' is short for labelled, 'unlab' for unlabelled, 'ord' for ordered, and 'unord' for unordered.

A different labelling of genes in Figure 3.13, for example, $(1, 4, 5, 2), (7)$, $(3, 8, 6)$, has the same probability as the original labelling $(1, 7, 5, 8), (3)$, $(4, 6, 2)$. Also a different ordering of $(n_1, n_2, n_3) = (4, 1, 3)$, for example, $(n_1, n_2, n_3) = (1, 3, 4)$, has the same probability as the original ordering. In Table 3.3, note that certain labelled, ordered configurations are very unlikely, namely those that divide the sample into k groups of approximately the same sizes. The number of orderings for a given series of $n_1, n_2, \ldots, n_k$, is

$$\frac{k!}{\lambda_1! \lambda_2! \ldots \lambda_K!}, \tag{3.15}$$

where K is the number of different integers among $n_1, n_2, \ldots, n_k$ and $\lambda_1, \lambda_2, \ldots, \lambda_K$ their frequency. For example, if $n_1 = 3, n_2 = 3$, and $n_3 = 2$ then $K = 2$ and $\lambda_1 = 2$ and $\lambda_2 = 1$, such that there are three different orderings of the integers 3, 3, and 2; namely $(3, 3, 2)$, $(3, 2, 3)$, and $(2, 3, 3)$. If the genes are labelled then there are

$$k! \tag{3.16}$$

possible orderings of the groups, because groups with the same number of genes now appear as distinct because of the labels.

The latter two probabilities are now obtained by multiplying equations (3.12) and (3.14) by equations (3.15) and (3.16), respectively.

For an unlabelled, unordered configuration the probability becomes

$$P_n(n_1, \ldots, n_k, \text{unllabelled, unordered}) = \frac{(n-k)!(k-1)!k!}{(n-1)!\lambda_1!\lambda_2!\cdots\lambda_K!}, \quad (3.17)$$

and finally, the probability of an unordered, labelled configuration is

$$P_n(n_1, \ldots, n_k, \text{labelled, unordered}) = \frac{(n-k)!(k-1)!n_1!n_2!\cdots n_k!k!}{n!(n-1)!}. \quad (3.18)$$

In the example discussed in Table 3.3, there are twenty-one ordered, unlabelled configurations, that each occurs with probability 1/21. The 21 ordered, unlabelled configurations fall into the five unordered, unlabelled configurations. The first, fourth, and fifth unordered configurations cover three ordered configurations, whereas the second and third cover six ordered configurations.

3.5.1 Unbalanced trees

Of particular interest is the basal split into two lineages at the root of the tree. The probability that this split results in the labelled, unordered partition $(i, n-i)$, $i = 1, 2, \ldots, \lfloor n/2 \rfloor$ ($\lfloor x \rfloor$ is the integer part of x, that is decimal points of x are removed), is

$$P_n(i, n-i, \text{labelled, unordered}) = \frac{2i!(n-i)!}{n!(n-1)}. \quad (3.19)$$

The probability for unlabelled genes is

$$P_n(i, n-i, \text{unlabelled, unordered}) = \frac{2 - \delta_{i,n-i}}{n-1}, \quad (3.20)$$

where $\delta_{i,n-i}$ is 1 if n is even and $i = n/2$, and otherwise 0. If $i = 1$ then the tree is said to be unbalanced and equation (3.19) becomes

$$P_n(1, n-1, \text{labelled, unordered}) = \frac{2}{n(n-1)}. \quad (3.21)$$

In large samples unbalanced trees are unlikely.

3.5.2 Example: Neanderthal sequences

Nordborg (1998) studied the tree of a combined sample of 986 human mitochondrial sequences and one Neanderthal sequence obtained by

Krings et al. (1997). The example is discussed in further detail in Chapter 8. Assuming random mating between Neanderthals and humans such an unbalanced tree is very unlikely, indeed from equation (3.21) it is just $2/(986 \cdot 985) = 2 \cdot 10^{-6}$. However, Nordborg pointed out that a large part of the human sample had found common ancestors at the time the sequenced Neanderthal lived (30,000–100,000 years ago). For example, if there were only five ancestors to the present human sample 30,000 years ago (and Nordborg found that plausible), the probability is $2/(5 \cdot 4) = 10\%$. Thus, in itself the topology does not provide strong evidence against interbreeding between Neanderthals and humans.

3.6 A single lineage

Consider a given gene in a sample of size n. How long does it take before this gene coalesces with another gene in the sample? How many events pass before it coalesces with another gene? To answer the first question consider the recursion

$$E(W_n) = \frac{2}{n}E(T_n) + \frac{n-2}{n}\{E(T_n) + E(W_{n-1})\}, \qquad (3.22)$$

where W_n denotes the time until the gene merges with another gene. The recursion has initial conditions $E(W_2) = E(T_2) = 1$ because in a sample of size two the gene merges with the other gene at the first and only coalescent event.

For a given n, the gene is involved in the first coalescent event with probability $2/n$. If this is so then $W_n = T_n$ which provides the first term on the right side of the equality sign. With probability $1 - 2/n$ the gene is not involved in the first coalescent event which implies that $W_n = T_n + W_{n-1}$; T_n plus the time W_{n-1} until the gene coalesces in a sample of size $n - 1$. That is, the second term.

Now $E(T_n) = 2/(n(n-1))$ and the recursion simplifies to

$$E(W_n) = \frac{2}{n(n-1)} + \frac{n-2}{n}E(W_{n-1}), \qquad (3.23)$$

which has solution

$$E(W_n) = \frac{2}{n}. \qquad (3.24)$$

Thus on average the time until a gene merges with another gene is inversely proportional to the sample size.

Similarly we can obtain the number of coalescent events in the sample's history until a given gene coalesces with another gene in the sample. Again fix a gene, say gene 1. The number of coalescent events, C_n, fulfils

$$E(C_n) = \frac{2}{n} + \frac{n-2}{n}\{1 + E(C_{n-1})\}. \qquad (3.25)$$

The reasoning here is similar to the reasoning underlying the recursion (3.22): $E(T_n)$ is replaced by 1 and W_n by C_n. The initial condition is $E(C_2) = 1$ because there is only one coalescent event in the history of a sample of size two. Solving the recursion gives

$$E(C_n) = \frac{n+1}{3}.$$

Thus about one-third of all coalescent events happen prior (going backwards in time) to the first event involving the given gene, but they all happen in the course of a short time span, namely $2/n$.

3.7   Disjoint subsamples

Another related problem is the following. Consider a sample of size n that is divided into two disjoint subsamples, A and B, of sizes k and $n-k$, respectively. This situation is depicted in Figure 3.14. In this section it is assumed that the partition is made on the basis of non-genetic information. This could, for example, be the geographic origin of the genes. A polymorphism partitioning the sample into two subsamples implies knowledge about (at least) one mutation event in the sample's history and this information changes the probabilistic structure of the tree: both the jump process and the waiting time process. This situation will be taken up in the subsequent section.

What is the probability that all genes in subsample A coalesce with each other before coalescing with any of the genes in B? This situation is distinct from the situation discussed by Saunders et al. (1984) because the genes in A are not allowed to merge with those of B. It is also distinct from the situation discussed in the section about hanging subtrees because we do not require that A and/or B subtend from a given set of fixed lineages.

Let the number of genes in A be k such that there are $n-k$ genes in B. The probability, $q_{k,n-k}$, that all genes in A find a MRCA before coalescing

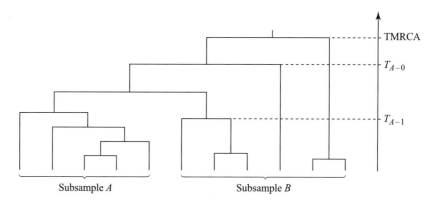

Figure 3.14   Illustration of division of the sample into disjoint subsamples A and B, where all the genes in subsample A coalesce before any of them coalesce with a gene in subsample B. T_{A-1} denotes the time of the MRCA of subsample A, T_{A-0} denotes the time of absorption of the last lineage of A into the genealogy of B.

with any gene in B can be calculated explicitly. It is given by

$$q_{k,n-k} = \frac{2(k-1)!}{(k+1)(n-1)\cdots(n-k+2)(n-k+1)}. \tag{3.26}$$

Similarly, the probability that B coalesces before coalescing with any of the genes in A is just $q_{n-k,k}$, that is, the roles of A and B are reversed. This implies that the probability, $r_{k,n-k}$, that one of the two samples finds a MRCA before coalescing with members of the other sample is

$$r_{k,n-k} = q_{k,n-k} + q_{n-k,k} - \frac{2k!(n-k)!}{n!(n-1)}, \tag{3.27}$$

where the last term is the probability that k and $n-k$ labelled genes subtend from the deepest split in the tree (see equation (3.19)). This term is subtracted because it is included in both other terms as illustrated in Figure 3.15.

The jump process can be described in a way similar to the description of the jump process of Saunders et al. (1984). If there are (i, j) ancestors of A and B, respectively, then

$$(i, j) \to (i-1, j) \text{ with probability } \frac{i+1}{i+j}, \tag{3.28}$$

and

$$(i, j) \to (i, j-1) \text{ with probability } \frac{j-1}{i+j}. \tag{3.29}$$

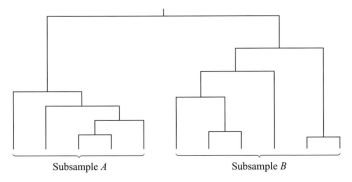

Subsample *A*   Subsample *B*

Figure 3.15   Illustration of division of the sample into disjoint subsamples *A* and *B*, such that all genes in *A* and *B* find MRCAs before the two subsamples find a MRCA. The probability of this is given in the last term of equation (3.27).

The process starts in $(k, n - k)$ and continues until $(1, j)$ for some j has been reached. From there it eventually jumps to $(0, j)$ for some j and further down to $(0, 1)$. Here 0 denotes that sample *A* has been fully absorbed into *B*.

To simulate the genealogy of *A* and *B* one might first simulate the jump process and then subsequently simulate the times between events. Since the jump process is independent of the times between events, these times are simply given by exponential distributions with intensities $\binom{i+j}{2}$, while there are $i + j$ genes in total.

Here three quantities are of interest, namely the time, T_{A-1}, of the MRCA of *A*, the time, T_{A-0}, until the last lineage of *A* is absorbed into the genealogy of *B*, and the number, ANC_B, of ancestors of *B* at the time of the MRCA of *A*. ($A - 1$ and $A - 0$ denote that there are 1 and 0, respectively, lineages of *A* in the sample, see Figure 3.14.) These quantities are key numbers in the description of the genealogy of the whole sample. One can give full probabilistic treatments of these quantities deriving the densities of all three. Here we shall only be concerned with their mean values.

The mean of T_{A-1} is

$$E(T_{A-1}) = \frac{k - 1}{n}. \qquad (3.30)$$

If k and $n - k$ are large then the mean is approximately equal to the frequency of *A* in the sample, and if $k = n - 1$ $(n - k = 1)$ then the mean is $(n - 2)/n$: The last coalescence event must involve the only gene in *B*, thus the time until a MRCA of *A* is the time from the present until there are two ancestors of the whole sample.

The mean of T_{A-0} is

$$E(T_{A-0}) = \frac{2k}{n}. \tag{3.31}$$

Note that $T_{A-0} - T_{A-1}$ is the time from the MRCA and until absorption. It has mean $(k+1)/n$, which is of the same order as $E(T_{A-1})$.

If k and $n - k$ are large, both the mean of T_{A-1} and $T_{A-0} - T_{A-1}$ equal approximately the frequency of A in the whole sample. In Section 3.8 we shall see how these times are affected when the split between A and B is caused by a mutation. Ignoring the mutation has notable effects, in particular if the mutant allele is found in low frequency, that is, if k/n is small.

Finally, we state the mean number, ANC_B, of ancestors of B at the time of the MRCA of A. Again, this number should be compared to the similar number conditioned on a mutation causing the split in the sample (Section 3.8). We find that

$$E(ANC_B) = \frac{3(n+1)}{k+2} - 2. \tag{3.32}$$

If both k and $n - k$ are large, $E(ANC_B) \approx 3/x - 2$, where $x = k/n$ is the frequency of A in the sample. Thus the number of ancestors of B is almost inversely proportional to the frequency of A.

3.7.1 Examples

Harris and Hey (1999) sequenced 4,200 bp of the PDHA1 gene in thirty-five male individuals from different human populations. Sixteen individuals were from African populations, and nineteen from non-African populations. The gene tree constructed after removing one incompatible site and rooted using a chimpanzee sequence is reproduced in Figure 3.16. Note that the MRCA is estimated to be almost 2 million years old. All the non-African sequences belong to haplotype A and B which are separated by just one mutation, whereas the African sequences contain much more diversity. One may ask whether this clustering of non-Africans in the gene tree provide evidence for restricted migration among African and non-African populations. Using the theory above, we may phrase the question as: What is the probability that a subsample A of size $k = 19$ coalesces before coalescing with an ancestor of the $n - k = 16$ remaining genes. Plugging into equation (3.26), the probability is $4.5 \cdot 10^{-11}$. Thus, it is highly unlikely that such a configuration would be observed in a randomly mating population.

We may also continue our Neanderthal example. Since the original investigation, two more Neanderthal individuals have been sequenced and

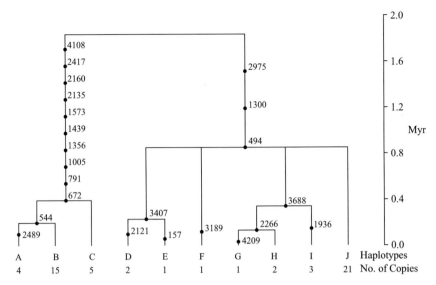

Figure 3.16   The gene tree of the PHDA1 gene from a sample of Africans and non-Africans. Haplotypes A and B were found in non-African population only, the remaining haplotypes in African populations only. The tree was rooted with a chimpanzee sequence. (Adapted from Harris and Hey 1999.)

found to cluster with the previous Neanderthal with the root of the tree separating humans and Neanderthals. Assume again that at the time of the Neanderthals there were five ancestors to the human sample. Then the probability of a split of eight lineages into three and five is given by the last term of equation (3.26), and is equal to $2 \cdot 3!(8 - 3)!/(8!(8 - 1)) = 0.5\%$. Thus, the evidence is much stronger than for one sequence. If we instead want to test the hypothesis that either all humans coalesced before coalescing with the Neanderthal sample or vice versa, the probability can be found from equation (3.26) to be 2.4%.

3.8   A sample partitioned by a mutation

Let us bring the discussion in the previous section one step further. Consider a sample of size n where a polymorphism divides the sample into two disjoint subsamples, A and B, of size k and $n - k$, respectively. Assume that just one mutation separates the two subsamples and that the scaled mutation rate is very low. This is often believed to be the case for many SNPs. Here the mutation rate for a single locus might be as low as $u = 10^{-8}$ per generation. If we put $N = 10^4$ then $\theta = 4Nu = 4 \cdot 10^{-4}$ which is almost zero.

Let us for now assume we know which allele is the oldest, say sample B carries the oldest allele. In order for the whole sample to find a MRCA sample A must find a MRCA before coalescing with any sequence in B and further, a mutation must have occurred on the branch connecting the genealogy of A with that of B (Figure 3.14). A mutation is more likely to happen on this branch if it is long. Remember that for a given mutation rate the number of mutations along a branch of length t is Poisson distributed with intensity $\theta t/2$ (see equation (2.6)). If θ is small then the probability of a single mutation is approximately $\theta t/2$, which increases linearly with the branch length. The probability of more than one mutation becomes negligible (of the order $(\theta t)^2$) and that of no mutations is $1 - \theta t/2$.

The branch connecting A with B tends to be longer if the number of ancestors of B at the time of the MRCA of A is small. That we know from Section 3.6: Here we found that the average time until one lineage was absorbed into the rest of the sample is $2/k$ if there are k sequences in total. Let $k - 1$ be the number of ancestors of B at the time of the MRCA of A such that there are k sequences in total. It transpires that the mutation has more time to occur if k is small.

The jump process is very similar to that given in equations (3.28) and (3.29). Let k and $n - k$ be the number of ancestors of A and B, respectively, and assume the current number of ancestors of A and B are i and j, respectively. Then

$$(i, j) \to (i - 1, j) \text{ with probability } \frac{i}{i + j - 1}, \tag{3.33}$$

and

$$(i, j) \to (i, j - 1) \text{ with probability } \frac{j - 1}{i + j - 1}. \tag{3.34}$$

The process starts in $(k, n - k)$ and continues until $(1, j)$ for some j has been reached. Eventually the state $(0, j)$ is reached where 0 denotes the occurrence of the mutation.

One can derive the expressions for the times until the MRCA of A (T_{A-1}), the time until the last lineage of A has been absorbed into the genealogy of B (T_{A-0}), and the number of ancestors of B at the time of the MRCA of A (ANC_B). In the previous section the similar quantities were discussed when the split between A and B was not conditioned on the split being caused by a mutation. Instead of T_{A-0} we will use T_{MUT}, the age of the mutation, which seems to be a more natural quantity in this context. Conditioned on a mutation the mean values become

$$E(T_{A-1} \mid \text{Mutation}) = \frac{2k}{n-1} - \frac{2}{n} - 2\binom{n-1}{k}^{-1} \sum_{i=2}^{n-k+1} \frac{1}{i}\binom{n-i-1}{k-2},$$

(3.35)

and

$$E(T_{\text{MUT}} \mid \text{Mutation}) = \frac{2k}{n-1} - \frac{2}{n} - 2\binom{n-1}{k}^{-1} \sum_{i=2}^{n-k+1} \frac{1}{i}\binom{n-i-1}{k-1}.$$

(3.36)

These expressions are fairly unintuitive and difficult to handle, but become somewhat more manageable if k and $n - k$ are large:

$$E(T_{A-1} \mid \text{Mutation}) \approx 2x + \frac{2x^2}{(1-x)^2}\{\log(x) + 1 - x\},$$

(3.37)

and

$$E(T_{\text{MUT}} \mid \text{Mutation}) \approx -\frac{2x}{1-x}\log(x),$$

(3.38)

where $x = k/n$. The number of ancestors of B at the time of the MRCA of A is approximately

$$E(\text{ANC}_B \mid \text{Mutation}) \approx \frac{2}{x}.$$

(3.39)

Figure 3.17 compares the mean values conditioned on a mutation to the values obtained without conditioning on a mutation causing the split between A and B.

3.8.1   Unknown ancestral state

If it is not known which of the two alleles is the oldest allele the situation is different from that described above. The probability that an allele found in frequency k out of n genes is the oldest is k/n. That is, the probability that A carries the mutant allele is $1 - k/n = (n - k)/n$. We can combine this probability with the jump process described above to obtain the transition probabilities of the jump process in the present case. The process makes

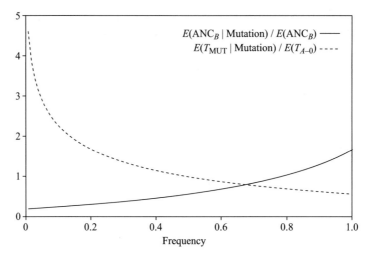

Figure 3.17 The dotted curve is the ratio of $E(T_{MUT} \mid \text{Mutation})$ to $E(T_{A-0})$. The solid curve is the ratio of $E(ANC_B \mid \text{Mutation})$ to $E(ANC_B)$. Since T_{MUT} and T_{A-0} do not exactly measure the same time the dashed line falls below 1 for large frequencies.

a jump from a state (i, j) to the state $(i - 1, j)$ with probability

$$\frac{j}{i+j} \cdot \frac{i}{i+j-1} + \frac{i}{i+j} \cdot \frac{i-1}{i+j-1} = \frac{i}{i+j}. \qquad (3.40)$$

Here $j/(i+j)$ is the probability that A carries the mutant allele and similarly $i/(i+j)$ is the probability that B carries the mutant allele. If B carries the mutant a jump to $(i - 1, j)$ is made with probability $(i - 1)/(i + j)$. Thus the transition probabilities are given by

$$(i, j) \to (i - 1, j) \text{ with probability } \frac{i}{i+j}, \qquad (3.41)$$

and

$$(i, j) \to (i, j - 1) \text{ with probability } \frac{j}{i+j}. \qquad (3.42)$$

Both i and j can become zero (but never both of them).

3.8.2   The age of the MRCA for two sequences

In the previous subsection it has been assumed that there was only one mutation in the sample and that the mutation rate was negligible. Here we consider the situation of two sequences with $S_2 = k$ segregating sites and want to evaluate the time until their MRCA, conditional on the k mutations. The density, $f(t \mid S_2 = k)$, of T_2 given $S_2 = k$ can be written in the form

$$f(t \mid S_2 = k) = \frac{P(S_2 = k \mid T_2 = t)}{P(S_2 = k)} f(t), \tag{3.43}$$

where $f(t)$ is the exponential density with rate one and S_2 conditional on $T_2 = t$ is Poisson with intensity θt. Putting the pieces together yields

$$P(S_2 = k) = \int_0^\infty \frac{(\theta t)^k}{k!} e^{-\theta t} e^{-t} \, dt = \frac{\theta^k}{(1+\theta)^{k+1}}, \tag{3.44}$$

and

$$f(t \mid S_2 = k) = \frac{(1+\theta)^{k+1}}{k!} t^k e^{-(1+\theta)t}, \tag{3.45}$$

such that T_2 conditional on $S_2 = k$ is Gamma distributed, $\Gamma(k+1, \theta+1)$. In particular the mean is

$$E(T_2 \mid S_2 = k) = \frac{k+1}{\theta+1}, \tag{3.46}$$

which is larger or smaller than the unconditional mean time, $E(T_2) = 1$, according to whether $k \geq \theta$ or $k \leq \theta$, respectively.

3.9   The probability of going from *n* ancestors to *k* ancestors

Despite the simplicity of the coalescent process, even answers to simple questions can have quite complicated expressions as answers. Such a question is the number of ancestors to j genes after a specified time period, t. The description so far only tells us the waiting time until the next coalescent event.

Figure 3.18 shows the probability of having 1, 2, 3, 4, 5, and 6 ancestral lineages when starting with seven sampled genes. Going back in time, the probability of the presence of all seven ancestors is almost one when t is very small, but then a pair will coalesce and there will only be six ancestors. This goes on until eventually there will be one ancestor to the whole sample. If $h_{n,k}(t)$ is the probability of k ancestral lineages after time t, when starting

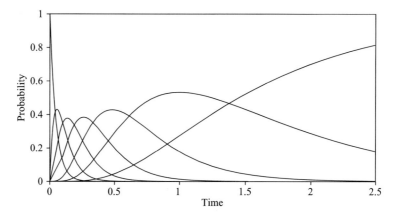

Figure 3.18 The probability of different numbers of ancestors starting with seven ancestors at time 0. The curve starting with probability 1 at time 0 and then descending towards 0, is the probability of seven ancestors. The curve that peaks first is the probability of six ancestors, then followed by the curve for having five ancestors, etc. The curve that starts in 0, but converges to 1 as time increases, is the probability of only one ancestor to the sample. At each time point the curves sum up to 1.

with n genes, then

$$h_{n,k}(t) = \sum_{i=k}^{n} e^{-\binom{i}{2}t} \frac{(2i-1)(-1)^{i-k} k_{(i-1)} n_{[i]}}{k!(i-k)! n_{(i)}}, \qquad (3.47)$$

where $n_{[i]} = n(n-1)\cdots(n-i+1)$ and $n_{(i)} = n(n+1)\cdots(n+i-1)$.

This probability can be obtained in a variety of ways. First, it is the solution to the sum of a series of exponential waiting times with different parameters. For there to be k ancestral lineages at time t to n sampled genes, $n-k$ coalescent events must have occurred, that is,

$$T_n + T_{n-1} + \cdots + T_{k+1} < t. \qquad (3.48)$$

However, this could allow too many coalescent events to have happened. So we must also impose the restriction that

$$T_n + T_{n-1} + \cdots + T_k > t. \qquad (3.49)$$

Another different, but illuminating, way to find $h_{n,k}(t)$ is to use the probability that a pair of genes have found a common ancestor or not after time t. All these $\binom{n}{2}$ probabilities are the same and can be combined into a statement about the probability that a sample of n has k ancestors at time t. For an illustration of three genes having two ancestors see Figure 3.19. Each circle is the event where the two genes have not found an ancestor after time t and has probability e^{-t}. The intersection of all three circles has probability

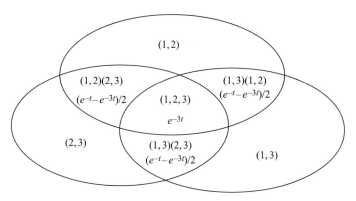

Figure 3.19 The probability that a sample of three genes have two ancestors at time t. An area is indicated with a set of sequences—these sequences have not found any common ancestors. The area with $1, 2, 3$ still have distinct ancestors after time t. It has probability e^{-3t}. Elementary calculations allow the probability of all events to be found.

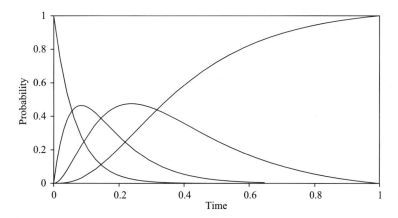

Figure 3.20 The probability of different numbers of ancestors starting with seven ancestors at time 0 and ending with four ancestors at different times. The curve that has probability 1 at time 0 is the probability of seven ancestors. The curve that has probability 1 at the right end of the interval is the probability of four ancestors. The curve that peaks first is the probability of six ancestors, followed by the curve for having five ancestors. At each time point the curves add up to one.

e^{-3t}, since it is the event that no genes have found common ancestors. The three intersections of two circles not including the third circle have probability $(e^{-t} - e^{-3t})/2$. Note that the figure is misleading because the part of a circle outside the intersections is empty: If two pairs of genes have found common ancestors at time t then all three pairs have. Therefore, all probabilities of the areas are defined and it is a simple addition–subtraction exercise to obtain probabilities of interest. The event where the three genes have two ancestors are the areas that are contained in two circles. The area outside any circle are the events where all have found a common ancestor.

Obviously, for large n and k book-keeping will be much more complicated, but would follow the same principles.

Since the coalescent process is Markovian, it is easy to find the probability of a descending number of ancestors to a series of increasing times. Here we will be concerned with the probability of a specific number of ancestors, say m, at time t_1 when the number of ancestors is known at the beginning and the end of the interval. First we need the probability that there are m ancestors at time t_1 and k at time t,

$$h_{n,m,k}(t_1,t) = h_{n,m}(t_1)h_{m,k}(t-t_1). \tag{3.50}$$

Thus, the probability of m ancestors at time t_1, when we started with n and ended with k at time t is

$$P(m \mid n,k;t_1,t) = \frac{h_{n,m}(t)h_{m,k}(t-t_1)}{h_{n,k}(t)}. \tag{3.51}$$

This is illustrated for $n=7$ and $k=4$ in Figure 3.20.

Recommended reading

Griffiths, R. C. and Tavaré, S. (1998), 'The age of a mutation in a general coalescent tree', *Stochastic Models* **14**, 273–295.

Saunders, I. W., Tavaré, S., and Watterson, G. (1984), 'On the genealogy of nested subsamples from a haploid population', *Adv. Appl. Probab.* **16**(3), 471–491.

Semple, C. and Steel, M. (2003), *Phylogenetics*, Oxford University Press.

Tavaré, S. (1984), 'Line-of-descent and genealogical processes, and their applications in population genetics models', *Theor. Popul. Biol.* **26**(2), 119–164.

4 Extensions to the basic coalescent

4.1 Introduction

The basic coalescent process is based on the Wright–Fisher model of reproduction, assuming constant population size, random mating, and absence of selection. These are very strict conditions, which are rarely, if ever, met in nature. Fortunately, the coalescent framework can easily be extended to include simple deviations from these basic assumptions, resulting in a process that is believed to model the true mating pattern more closely. In what follows, we consider in turn how to deal with the following scenarios: fluctuating population size, population structure, and different kinds of selection. It should be stressed that it is relatively easy to combine these extensions in a single model if necessary.

The focus here is to provide means to simulate the distribution of the genealogy of a sample of genes and the distribution of the sample configuration. Further, it is our interest to point out some major differences between genealogies generated under the basic coalescent and under the scenarios discussed here. In Chapter 2 we learned how a sample configuration under neutrality can be generated in two steps: First simulate the genealogy, then add mutations to the genealogy. In the scenarios involving selection the genealogy and the sample configuration can be simulated at the selected site. Subsequently, neutral (intra-allelic) mutations can be added to the genealogy in a second step, similarly to the procedure just mentioned above.

Throughout the chapter we assume sequences are non-recombining. The effects of recombination and other forms of genetic exchange between chromosomes are considered in Chapter 5.

4.2 The coalescent with fluctuating population size

4.2.1 Stochastic and systematic changes

Populations commonly fluctuate in size over time. Fluctuations are, for example, caused by variation in offspring number or by changes over time in the opportunities for the species. Changes can either be extrinsic (due to the environment) or intrinsic (due to the competitive ability of the species).

From a modelling perspective, we distinguish between stochastic and systematic changes. Systematic changes are trends over years in the population size to change in a certain way. The most prominent example is the human population, which has experienced a monotonic increase over an extended period of time (see Table 1.2). Experimental evidence suggests that this period has been sufficiently long such that the population growth should be incorporated into a proper analysis of human Y-chromosome and mtDNA sequence data. Nuclear DNA have a larger effective population size because it is inherited through both the female and male germ line in two copies. The coalescent time scale is therefore larger, and some data sets appear to display variability predating the period of rapid growth of the human population. Thus, for these data one may need a combined model of constant (or nearly constant) population size and population size expansion to account for the observations (see below). Many animals associated with humans and several emergent viruses have also experienced a steady increase in population size. It is more difficult to find good examples of species with a steady decline except for species affected by humans in the very recent past, but such declines are too recent to be detectable in sequence data. The only exception may again be certain viruses, where vaccination campaigns and other measures have decreased effective population sizes dramatically (e.g. polio or measles).

Systematic trends are modelled deterministically such that the population size at time t is given by $N(t)$, a function of t only. In addition, a stochastic term might be added to $N(t)$ to reflect random deviations from $N(t)$ caused by environmental (e.g. draught) or other factors. Random deviations from the trend (i.e. from $N(t)$) are assumed to be of minor importance compared to the trend itself. These deviations might therefore be ignored. This is also the first assumption we make below.

4.2.2   How to model population changes in the coalescent

In order to simplify the mathematical exposition considerably we will make some convenient assumptions. First of all we consider only changes that are fully deterministic in nature, that is the (effective) size of the population is at any point, t, in time given as a function of t, namely $N(t)$. In the following $N = N(0)$.

Recall that the natural time scale of the basic coalescent process is in units of $2N$ generations, that is one generation counts $1/(2N)$ units in the continuous time analogue. This is also the probability by which two genes find a common ancestor in the previous generation in the discrete time coalescent. If the population size changes over time things are different. The probability that two genes coalesce in the previous generation is $p(t) = 1/(2N(t))$ which might differ from $p(0)$, and the natural scale of the coalescent process is $2N(t)$ locally. To simulate and describe genealogies probabilistically under scenarios with changing population sizes, changes in $p(t)$ have to be taken into account. When $N(t)$ is smaller than N, for example, if the size of the population is declining as we move backwards in time, the probability of a coalescence event increases and a MRCA is found more rapidly than if $N(t)$ is constant over time.

One way to take this into account is to generate genealogies under the constant size coalescent process and then to stretch or compress time, according to whether $p(t)$ is smaller or larger than $p(0)$. If $p(t)$ is smaller than $p(0)$ by a factor of five (for example) then time should be stretched locally by a factor of five to accommodate this. Figure 4.1 illustrates this. The figure can be read either bottom-up or top-down: In the bottom-up perspective real time (in generations or some units of generations) is transferred into an equivalent amount of time in the constant size coalescent process, and in the top-down perspective, time in the constant size coalescent is transferred into an equivalent amount of real generation time.

The second convenient assumption we will make is that (1) $N(t)$ is given in terms of continuous time (in units of $2N$ generations) and that (2) $N(t)$ need not be an integer. (2) is just to allow for some flexibility in specification

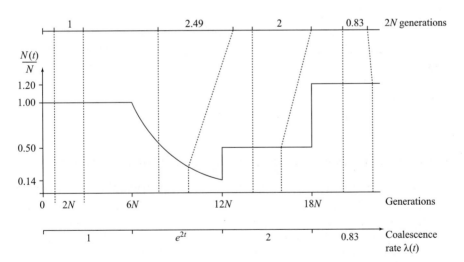

Figure 4.1   Stretching and compressing time in the coalescent process. The figure shows how the size of an example population changes over time (in generations). When the size decreases the rate of coalescence for two genes to find a common ancestor increases. Each of the intervals between dashed lines represents $2N$ generations. The top line shows how much time these intervals account for in the basic coalescent. The middle line shows time in the example population and the lower line shows the coalescence rate in the example population.

of $N(t)$ and can easily be dispensed with. Define

$$\Lambda(t) = \int_0^t \frac{1}{\lambda(u)}\, du, \tag{4.1}$$

where $\lambda(t) = N(t)/N$, the relative size of $N(t)$ to N. $\Lambda(t)$ is the accumulated coalescent rate over time measured relative to the rate at time $t = 0$, that is $\lambda(0) = 1$. $\Lambda(t)$ is also the harmonic mean of the relative population sizes and provides the means to stretch and compress time. To ensure that a sample of genes always has a MRCA, $\Lambda(\infty)$ must be ∞.

Let $T_2, \ldots, T_n$ be the waiting times while there are $2, \ldots, n$ ancestors of the sample and let $V_k = T_n + \cdots + T_k$ be the accumulated waiting times from there are n genes until there are $k - 1$ ancestors. The distribution of T_k conditional on $V_{k+1} = v_{k+1}$ is

$$P(T_k > t \mid V_{k+1} = v_{k+1}) = \exp\left\{ -\binom{k}{2} \left(\Lambda(t + v_{k+1}) - \Lambda(v_{k+1}) \right) \right\}, \tag{4.2}$$

where $v_{n+1} = 0$. Note that if $\Lambda(t)$ in equation (4.2) is replaced by $\Lambda(t) = t$ the exponential density reappears. Generally, we have to keep track of when the last coalescent event occurred (V_{k+1}) in order to simulate the time of the next event (T_k), but none of the times of earlier events are required. This is an instance of the Markov property: the waiting time until the next event depends only on the time of the previous event.

To be able to simulate T_k we need to distinguish between times in the basic coalescent process and times in the variable population size coalescent. Henceforth, in this section, time in the basic coalescent is denoted by T_k^*. An algorithm to simulate T_k is the following:

Algorithm 4

1. Simulate $T_2^*, \ldots, T_n^*$ according to the basic coalescent, where T_k^* is exponentially distributed with parameter $\binom{k}{2}$. Denote the simulated values by t_k^*.

2. Solve $\Lambda(t_k + v_{k+1}) - \Lambda(v_{k+1}) = t_k^*$ for v_k, $k = 2, \ldots, n$, and $v_{n+1} = 0$.

3. The values $t_k = v_k - v_{k+1}$ are an outcome of the process, $T_2, \ldots, T_n$, described in equation (4.2).

In some cases the equation for v_k can be solved explicitly, in other cases one must resort to numerical calculations. The next section presents one case for which the solution is straightforward. Mutations are added to the genealogy using Algorithm 3.

4.3 Exponential growth

The simplest (and most natural) model of steady population growth is exponential growth, because it assumes that the instantaneous growth rate is proportional to the current population size,

$$N(t) = Ne^{-\beta t} \qquad (4.3)$$

where $\beta = 2Nb$ is the scaled growth rate. The minus sign preceding β signals that we are measuring time retrospectively, so in coalescent terms we are modelling an exponentially growing population as an exponentially declining population back in time. In this case $\lambda(t) = e^{-\beta t}$,

$$\Lambda(t) = \frac{1}{\beta}(e^{\beta t} - 1) \qquad (4.4)$$

and

$$P(T_k > t \mid V_{k+1} = v_{k+1}) = \exp\left\{-\binom{k}{2}\frac{1}{\beta}e^{\beta v_{k+1}}\left(e^{\beta t} - 1\right)\right\}. \qquad (4.5)$$

It follows from Algorithm 4 that t_k defined by

$$t_k = \frac{1}{\beta}\log(1 + \beta t_k^* e^{-\beta v_{k+1}}), \qquad (4.6)$$

$k = 2, \ldots, n$, is an outcome of the coalescent with exponential growth, β. The t_ks are found explicitly from the t_k^*s and the previous values $t_{k+1}, \ldots, t_n$ through $v_{k+1} = \sum_{i=k+1}^{n} t_i$.

The waiting times $T_2, \ldots, T_n$ are no longer independent of each other as in the basic coalescent but are negatively correlated: If one of them is large the others are more likely to be small, because the variation in the time until the MRCA is much smaller than for the basic coalescent. This is shown in Table 4.1.

Table 4.1   Correlation between waiting times[a]

n, β	0	10	100	1000
10	0	−0.26	−0.72	−0.87
100	0	−0.01	−0.04	−0.28

[a] Table shows the correlation between the time until the first coalescence event, T_n, and the time from T_n until the time of the MRCA, $\sum_{i=2}^{n-1} T_i$, for various growth rates, β, and sample sizes, n. A total of 10^5 simulations were performed for each combination of parameter values, except for $\beta = 0$ where analytical results are known.

4.3.1   The genealogy under exponential growth

The effect of exponential growth on the coalescent tree can be illustrated by simulating the coalescent of a sample of ten genes. In Figure 4.2 there is no population growth and the large variation in the height of the genealogy and the long upper branches is apparent. With exponential growth at a rate of $\beta = 1000$, the genealogies have very long terminal branches and much less variation in TMRCA (Figure 4.3). The smaller variation in TMRCA is caused by the deterministic decrease in population size. At time t the coalescent rate is $\lambda(t)\binom{k}{2} = e^{\beta t}\binom{k}{2}$, which should be compared to the constant rate in the basic coalescent, $\binom{k}{2}$. Coalescence events happen at an exponentially increasing rate with t.

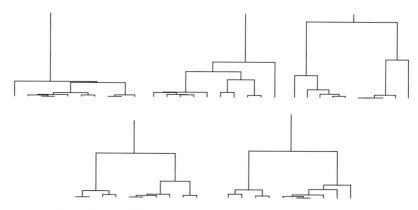

Figure 4.2   Five replicates of the coalescent process with constant population size for a sample of ten genes. Note the large variance in the time of the MRCA among replicates.

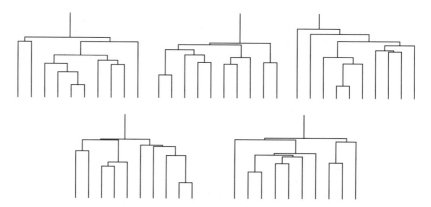

Figure 4.3   Five replicates of the coalescent with exponential growth, $\beta = 1000$, for a sample of $n = 10$ genes. Note the smaller variance in the time until the MRCA compared to the same quantity in Figure 4.2.

This implies that a typical coalescent genealogy under exponential growth has relatively shorter branches closer to the root than a genealogy without growth, or equivalently relatively longer terminal branches (Table 4.2). With high levels of exponential growth, the tree becomes almost star-shaped, and the sampled sequences are expected to be equally diverged from each other. It is important to realise that the degree of star-shape depends strongly on the sample size. In Table 4.2 it is measured by the ratio of the mean of the length of terminal branches to the mean of the total length of the tree. With growing sample size this measure of star-shape decreases. Figure 4.4 shows the distribution of the length of terminal branches to that of the whole tree. For small sample sizes it can be difficult to reject a null hypothesis of no growth, whereas for larger sample sizes the power to reject the null hypothesis is higher. This is a general feature of the models falling under the framework introduced in Section 4.2.2.

The differences in distribution between no growth and growth can also be seen in the distribution of pairwise differences between all pairs of sequences, also termed the mismatch distribution. With increasing rate of exponential growth, the mismatch distribution becomes unimodal, in contrast to the basic coalescent where the mismatch distribution typically is bi- or multimodal due to one of more deep splits in the genealogy caused by waiting a long time for the last lineages to coalesce. Figures 4.5 and 4.6 show typical mismatch distribution for simulated data sets evolving under constant population size and exponential growth, respectively.

From a data analysis angle, the long terminal branches of coalescent genealogies under exponential growth should be reflected in an excess of singleton polymorphisms in the frequency spectrum. Figure 4.7 shows the spectrum for four different values of β.

Table 4.2 The ratio of the mean of the length of terminal branches to the mean of the total length of the tree[a]

n, β	0	5	10	100	1000	10,000
10	0.35	0.54	0.59	0.74	0.83	0.88
100	0.19	0.33	0.37	0.56	0.72	0.82
1000	0.13	0.21	0.24	0.37	0.56	0.72
10,000	0.10	0.14	0.16	0.24	0.37	0.56

[a] For β increasing and n fixed the ratio increases towards one (perfect star-shape). For increasing n and fixed β the ratio decreases and the star-shaped form of the tree disappears eventually. For each value of $\beta > 0$, 10^5 replicates were used for $n = 10$ and $n = 100$, 10^4 for $n = 1000$, and 10^3 for $n = 10,000$. For $\beta = 0$ exact results are known.

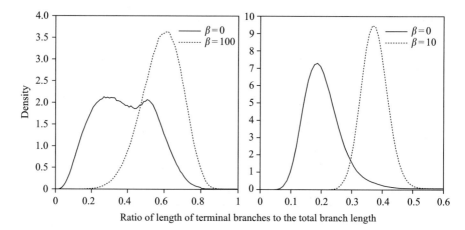

Figure 4.4   The distribution of the ratio of the length of terminal branches to that of the whole tree. In the left plot the distribution is shown for $n = 10$ and $\beta = 0, 100$. In the right plot it is shown for $n = 100$ and $\beta = 0, 10$. The power to reject the null hypothesis is 35.8% in the first case, 96.5% in the second. The figure is based on 10^6 replicate simulations for each set of parameters.

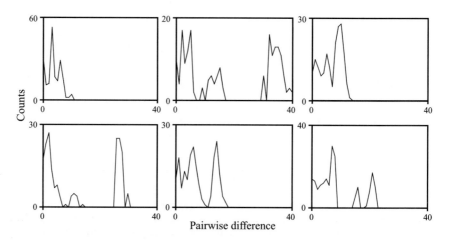

Figure 4.5   Mismatch distributions for six replicate simulations: constant population size. Results based on twenty sampled sequences with $\theta = 10$. The average pairwise number of differences is $\theta = 10$.

The variance in coalescent times changes greatly under exponential growth compared to the basic coalescent. In the basic coalescent, the last coalescent event, T_2, contributes 50–56% to the mean time until the MRCA if $n \geq 10$, but at least 86% to the variance of the same variable. Under exponential growth, there is not a closed expression for the mean and variance of the kth waiting time. Table 4.3 shows the mean and variance of the waiting

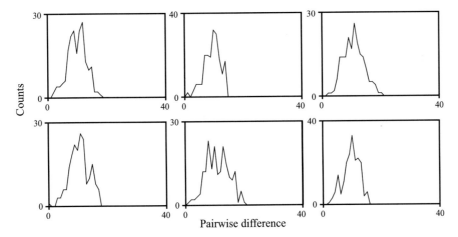

Figure 4.6   Mismatch distributions for six replicate simulations: exponential growth, $\beta = 1000$. Results based on twenty sampled sequences with $\theta = 158$, yielding an average pairwise number of differences of ten, similar to that of Figure 4.5.

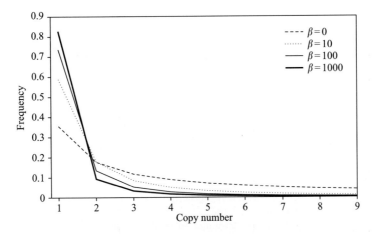

Figure 4.7   Frequency spectrum for $\beta = 0, 10, 100$, and 1000, and $n = 10$. Mutants occurring only once in a sample become increasingly more frequent with increasing β. Even for $\beta = 10$ there is more than 50% chance that a mutation is found in only one copy.

times calculated for the basic coalescent and estimated using simulation for three different rates of exponential growth, $\beta = 10, 100$, and 1000. For comparison, it is generally believed that recent human population growth is closest to $\beta = 1000$. Table 4.3 shows that with increasing population growth, the upper branches in the genealogy contribute less to both the mean coalescent time and to its variance.

In absolute values means and variances of coalescence times also differ from those under the basic coalescent. Table 4.4 provides a few examples.

Table 4.3   The percentage of the mean, $E(T_k)/\sum_{i=2}^{n} E(T_i)$, and the variance, $\text{Var}(T_k)/\sum_{i=2}^{n} \text{Var}(T_i)$, attributable to each coalescent event for four different rates of population growth and $n = 10$

k	$\beta = 0$		$\beta = 10$		$\beta = 100$		$\beta = 1000$	
	Mean	Var	Mean	Var	Mean	Var	Mean	Var
2	55.6	86.3	24.8	37.4	15.1	17.2	10.5	9.9
3	18.5	9.6	15.7	18.1	10.2	9.7	7.2	5.7
4	9.3	2.4	12.0	11.6	8.5	7.3	6.0	4.4
5	5.6	0.9	10.0	8.4	7.8	6.5	5.6	4.1
6	3.7	0.4	8.7	6.6	7.8	6.7	5.8	4.5
7	2.6	0.2	7.9	5.4	8.4	7.8	6.5	5.9
8	2.0	0.1	7.3	4.6	9.8	10.0	8.3	9.6
9	1.5	0.1	6.9	4.1	12.8	14.7	13.4	22.1
10	1.2	0.0	6.6	3.7	19.5	20.0	36.6	33.7

Note: $\sum_{j=2}^{n} \text{Var}(T_j)$ is not the variance of the time until a MRCA (unless $\beta = 0$), because the T_is are dependent variables. Results are based on 10^5 replicates for each growth rate, except for $\beta = 0$, where analytical expressions are known.

Table 4.4   The mean and variance of the TMRCA in a sample of size $n = 10$ for various values of β^a

β	Mean	Variance	SD/Mean
0	1.80	1.16	0.59
10	$2.80 \cdot 10^{-1}$	$2.80 \cdot 10^{-3}$	0.19
100	$5.04 \cdot 10^{-2}$	$3.18 \cdot 10^{-5}$	0.11
1000	$7.33 \cdot 10^{-3}$	$3.23 \cdot 10^{-7}$	0.08

[a] Also shown is the ratio of the standard deviation (SD), the square root of the variance, to the mean. A total of 10^6 replicates were obtained for all values of β, except for $\beta = 0$ where analytical expressions are known.

4.4   Population bottlenecks

A severe but short-lasting decline in effective population size is termed a population 'bottleneck'. Bottlenecks are normally associated with external catastrophic events such as an ice age or severe disease, but they can also be associated with the colonisation of a new habitat by a species. This could be colonisation of a new island by a plant species or the infection of a new host by one or a few virus particles. Population bottlenecks may also be associated with speciation and domestication of species, a well-studied

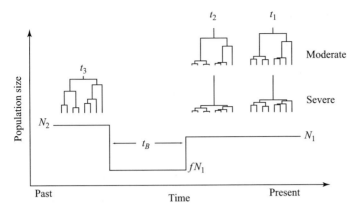

Figure 4.8 The effect of a population bottleneck on the genealogical tree of a sample. The shape of the genealogical tree depends on when the genes are sampled (or equivalently when the bottleneck occurred relative to the time of sampling). In the figure example trees are shown when sampling occurs after the bottleneck, shortly before the bottleneck and some time before the bottleneck (time is running backwards). f denotes the fraction of the current population surviving through the bottleneck and t_B is the duration of the bottleneck. Severe bottleneck: most coalescence events occur during the bottleneck. Moderate bottleneck: some lineages are likely to survive through the bottleneck and find MRCAs while the population has size N_2.

example being the population bottleneck associated with the domestication of maize (e.g. Eyre-Walker et al. 1998).

During a bottleneck, the rate of coalescence is increased, because the chance of sampling the same parental gene twice is increased when the effective population size is small. A simple model of a bottleneck is shown in Figure 4.8. The effective population size before the bottleneck (measuring time retrospectively) is N_1 and the population size after the bottleneck is N_2. The model has two additional parameters, the fraction of the population alive during the bottleneck (f), and the duration of the bottleneck (t_B). Whereas an instantaneous increase in effective population size can happen (catastrophic events), instantaneous drop is an oversimplification that is sometimes replaced by models of rapid population growth, such as the model of exponential growth discussed in the previous section. The probability that a pair of chromosomes coalesces during the bottleneck is $1 - \exp(-t_B/f)$, since the probability of no coalescence is exponentially distributed with parameter $1/f$ (where time is measured in units of $2N_1$ generations). Thus, the strength of a bottleneck is mainly determined by the product of its length (t_B) and severity ($1/f$).

Genealogies under bottlenecks can easily be simulated using Algorithm 4. Simulate according to the basic coalescent until the start of the bottleneck. From there on, and until the end of the bottleneck, simulate times from the basic coalescent with all times shortened by a factor of f. At the end of the bottleneck simulate again according to the basic coalescent, this time with all coalescent times multiplied by a factor of $f' = N_2/N_1$. Mutations can be added to the genealogy according to Algorithm 3.

4.4.1 Genealogical effect of bottlenecks

The effect of a population bottleneck on a sample depends on the strength of the bottleneck and on how long ago it occurred. Figure 4.8 shows how genealogies of a sample of genes taken at different time points may look for a severe or a moderate bottleneck.

A severe bottleneck, Figure 4.8 lower panel: If sequences are sampled just before the bottleneck (time point t_2, Figure 4.8), it is likely that all genes will coalesce during the bottleneck, and the genealogy will be very shallow. Sequences would then be expected to be (almost) identical. If sequences were sampled a while before the bottleneck (time point t_1) the genealogy would be deeper but most coalescent events would occur at the same time (namely, the time of the bottleneck). Thus, the genealogy will have relatively long external branches. This situation is similar to the situation with exponential growth, with an excess of singleton mutations, and Tajima's D (see Chapter 2) is expected to be negative.

A moderate bottleneck, Figure 4.8 upper panel: In this case not all lineages would be expected to coalesce during the bottleneck. For a set of genes sampled just before the bottleneck (time point t_2), most coalescence events would happen in a short time during the bottleneck, whereas the lineages surviving the bottleneck will take much longer to coalesce, since their coalescence rate is determined by the population size N_2 after the bottleneck. This genealogy will be very similar to what is expected for a subdivided population (see Section 4.6). In data this will show as an excess of intermediate frequency variants, causing Tajima's D to be positive. For a sample at time point t_1, the genealogy will show some long external branches, with a burst of coalescence events corresponding to the time of the bottleneck. This might be detectable as an excess of both singleton polymorphisms and of intermediate frequency variants in the data. In this situation, a single summary statistic such as Tajima's D would thus not be sufficient to summarise the deviation from the basic coalescent model. These conclusions are tentative and depend on the actual numbers n, N_1, f, t_1, and t_2.

For a sample taken after the bottleneck (time point t_3), the basic coalescent reappears with time scale in $2N_2$ generations.

The bottleneck model is sometimes used for studies of nuclear sequence data from human populations, with the bottleneck being associated with human migration out of Africa.

4.5 Effective population size revisited

Assume a model of an actual physical population is given, and let T_2 be the time until a MRCA of two randomly chosen genes. The effective population size was defined in Section 1.10 as

$$N_e^{(t)} = \frac{E(T_2)}{2}, \tag{4.7}$$

(T_2 is here in units of generations). This implies that two genes on average differ by the same number of mutations as a pair of genes in a Wright–Fisher model of constant size $N_e^{(t)}$, if $\theta = 4N_e^{(t)}u$ in both cases. $N_e^{(t)}$ depends on the actual size of the population as well as mating structure and offspring distribution. In Chapter 1 we gave examples where $N_e^{(t)}$ explicitly could be calculated. For the models introduced in this chapter $N_e^{(t)}$ can only be found by simulation. Table 4.5 provides some examples. The table also illustrates that the effective population size does not capture the structure of the population completely: The ratio of the standard deviation to the mean of T_2 differs for all values of β.

One way of determining the effective population size for a real physical population is to calculate the average pairwise difference $\hat{\pi}$ between genes in a sample. If an estimate, $\hat{u}$, of the mutation rate per generation is known (e.g. from experimental investigations), then an estimate of $N^{(t)}$ is

$$\frac{\hat{\pi}}{4\hat{u}}, \tag{4.8}$$

because $E(\hat{\pi}) = \theta = 4N_e^{(t)}u$.

Table 4.5 The effective population size for different values of β assuming $N(0) = 1{,}000^a$

β	0	10	100	1,000
$N_e^{(t)}$	1,000	200	41	6
$N(0)$	1,000	5,000	24,400	159,000
SD/Mean	1.00	0.45	0.29	0.063

[a] The second line shows what the initial size should be if the effective size is 1,000 for all values of β. The bottom line shows the ratio of the standard deviation (SD), the square root of the variance, to the mean of T_2.

4.6 The coalescent with population structure

The basic coalescent and the coalescent with fluctuating population size assume that all genes are exchangeable, which implies that at a given coalescent event, each possible pair of genes are equally likely to coalesce. An obvious case where this assumption is not met is for populations with mating structure. In this case individuals tend to mate with other individuals in physical (or social) proximity. Most, if not all, species display population structure to some degree. In humans, the probability of mating within human races or continents is greater than that of mating between. In a large rape seed field, the probability of pollination is larger for individual plants close together than for individual plants at opposite ends of the field.

Several models of population structure have been proposed and analysed. The mathematically most tractable models are deme models, where the population is divided into a number of units called demes. Within each of these random mating prevails, but between demes there is limited exchange of gametes through migration. Another class of models assumes no discrete random mating units, but focuses on a continuous space representation where genes are located and move around.

Coalescent theory has mainly focused on deme models, which will be dealt with in detail; a recent continuous space model will be briefly discussed at the end of this section.

4.6.1 The finite island model

Assume that a population is divided into d islands (also termed demes) of equal size $2N$ (N diploids or $2N$ haploid individuals), with a total population size of $2Nd$ genes. Each island contributes with a fraction m to a pool of migrants from which the island in return receives an equal proportion of migrants. Thus, all demes can be treated equally.

Assume n_i genes are sampled from deme i, that is, $n = \sum_{i=1}^{d} n_i$ is the total number of genes sampled. The coalescent process now needs to take the location of genes into account since genes can only coalesce if they are in the same deme, and genes can only get to the same deme by migration. If we assume $2Nd$ is large we can again use a continuous time approximation to the coalescent process with time scaled in units of $2Nd$ generations. The time until the first coalescent event is exponentially distributed with parameter

$$I_{\text{coal}} = d \sum_{i=1}^{d} \frac{k_i (k_i - 1)}{2},$$
(4.9)

while there are k_i ancestors located in deme i. The k_i ancestors in deme i need not be ancestors of the n_i genes initially sampled in deme i, because of migration between the demes. Note the factor d in front of the sum. Since time is measured in units of $2Nd$ (and not $2N$) generations there is a d preceding the rate $\binom{k_i}{2}$. In addition, migration events happen at rate

$$I_{\mathrm{migr}} = \frac{Mdk}{2}, \tag{4.10}$$

where M is the scaled (or population) migration rate, $M = 4Nm$ (scaled in the size of a single deme), m is the proportion of migrants joining and leaving a deme in each generation, and $k = \sum_{i=1}^{d} k_i$ the total number of ancestors of all demes. The argument for the rate I_{migr} is similar to the argument given for the rate until a mutation event occurs (with M replaced by θ, Section 2.2).

In total this leads to a rate of $I_{\mathrm{coal}} + I_{\mathrm{migr}}$ until the first event. This event is a migration event with probability

$$\frac{I_{\mathrm{migr}}}{I_{\mathrm{migr}} + I_{\mathrm{coal}}} = \frac{\sum_{i=1}^{d} k_i(k_i-1)}{kM + \sum_{i=1}^{d} k_i(k_i-1)} = \frac{\sum_{i=1}^{d} k_i^2 - k}{k(M-1) + \sum_{i=1}^{d} k_i^2}, \tag{4.11}$$

and a coalescence event with probability

$$\frac{I_{\mathrm{coal}}}{I_{\mathrm{migr}} + I_{\mathrm{coal}}} = 1 - \frac{I_{\mathrm{migr}}}{I_{\mathrm{migr}} + I_{\mathrm{coal}}} = \frac{kM}{k(M-1) + \sum_{i=1}^{d} k_i^2}. \tag{4.12}$$

If a migration event occurs, a gene is chosen at random and moved to a deme also chosen at random. If a coalescence event occurs a deme is chosen weighed according to the coalescence rates, that is, deme j is chosen with probability

$$\frac{k_j(k_j-1)}{\sum_{i=1}^{d} k_i(k_i-1)}, \tag{4.13}$$

and two genes within that deme are chosen randomly to coalesce. (Note that demes with $k_i = 1$ cannot be chosen.) An example of a genealogy is shown in Figure 4.9.

These considerations provide an algorithm for simulating the coalescent with population structure. The algorithm is similar to Algorithm 2; each time an event happens the number of ancestors is updated according to the event type. It stops when there is only one ancestor of the whole sample.

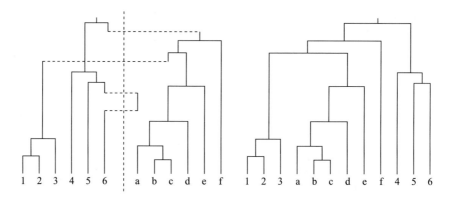

Figure 4.9   The coalescent with migration in two demes. Six genes sampled from each deme. The left panel shows migration events and coalescent events (associated with vertical lines), the right panel just the resulting coalescent tree.

4.6.2   The coalescent tree in the finite island model

In the simple finite island model, the mean and variance of coalescent times for a sample of genes can be calculated. Obviously, both are dependent on how the n genes are sampled, that is, whether they are from the same or from different demes. Let the sample configuration be given as $\mathbf{d} = (d_1, d_2, \ldots, d_n)$ where d_i is the number of demes with i members. (This notation is very similar to the notation introduced in Section 2.4.1 to describe the sample configuration in the infinite alleles model.) In the beginning,

$$n = \sum_{i=1}^{n} i\, d_i. \tag{4.14}$$

Subsequent events change $\mathbf{d}$ until the state $(1, 0, \ldots, 0)$ is reached. Let $T(\mathbf{d})$ denote the time until the MRCA given the sample configuration is $\mathbf{d}$, in particular $T(2, 0)$ is the time for two genes in two different demes to find a MRCA, and $T(0, 1)$ is the time for two genes in the same deme to find a MRCA. We will find the expectation and variance of $T(2, 0)$ and $T(0, 1)$ using simple recursions analogous to those introduced in previous sections.

Two genes cannot find a MRCA unless they are in the same deme, and they must do so before one of them migrates. Thus,

$$E(T(0, 1)) = \frac{1}{Md + d} + \frac{Md}{Md + d} E(T(2, 0)) \tag{4.15}$$

and

$$E(T(2, 0)) = \frac{d - 1}{Md} + E(T(0, 1)), \tag{4.16}$$

which can be solved to give

$$E(T(0,1)) = 1, \tag{4.17}$$

and

$$E(T(2,0)) = 1 + \frac{d-1}{Md}. \tag{4.18}$$

Likewise, the recursions for the variance of coalescence times for a sample of two genes can be solved to give

$$\text{Var}(T(0,1)) = 1 + \frac{2(d-1)^2}{Md^2}, \tag{4.19}$$

and

$$\text{Var}(T(2,0)) = 1 + \frac{2(d-1)^2}{Md^2} + \frac{(d-1)^2}{M^2 d^2}. \tag{4.20}$$

Equations (4.17)–(4.20) might differ from those of other authors because of differences in scaling of time and definition of M. The variances of $T(0,1)$ and $T(2,0)$ increase with increasing subdivision (decreasing M), as illustrated in Table 4.6.

It is quite surprising that the expected time until the MRCA of two genes from the same deme is independent of the migration rate. Intuitively it may be understood as follows. When migration is decreased, the expected TMRCA for two genes within the same deme decreases conditional on the event that none of the genes migrate before coalescing. However, if one (or both) migrate, then the expected TMRCA increases with decreasing migration rate because the waiting time until the genes meet in the same deme again increases. These two effects exactly balance to give a mean

Table 4.6 The mean and variance of the time until a MRCA for two genes as a function of the migration rate[a]

M	∞	10	1	0.1	0.01
$E(T(0,1))$	1	1	1	1	1
$E(T(2,0))$	1	1.08	1.75	8.5	76
$\text{Var}(T(0,1))$	1	1.11	2.13	12.3	114
$\text{Var}(T(2,0))$	1	1.12	2.69	68.5	5740

[a] There are four equally sized demes. If $M=\infty$, $T(0,1)$ and $T(2,0)$ are exponentially distributed with intensity 1.

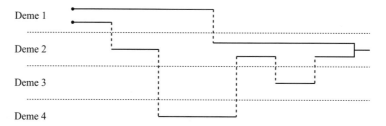

Figure 4.10 Jumping between demes before a MRCA is found.

time independent of the migration rate. The argument also illustrates that the variance of $T(2,0)$ and $T(0,1)$ depends on M: When M decreases, the variance increases as is apparent from Table 4.6. Thus, even though the means of the two distributions are identical and independent of the migration rate, the distributions are very different.

We can gain further intuition about the process if it is studied from a different perspective. Again assume a sample of two genes. The genes' lineages jump between being in the same deme and being in different demes until the lineages coalesce, as in Figure 4.10. Assume the two genes initially are located in different demes. First, they have to migrate into the same deme. In that particular deme the genes coalesce with probability $1/(M+1)$ and one of them leaves the deme with probability $M/(M+1)$. If one of them leaves they eventually end up in the same deme again, and the process repeats itself. The probability that this pattern (different demes, same deme) happens j times before the MRCA is a geometric probability, namely

$$P(j \text{ migration events}) = \frac{1}{M+1}\left(\frac{M}{M+1}\right)^{j-1}, \qquad (4.21)$$

$j = 1, 2, \ldots$, and the two genes spend j epochs in the same deme, j epochs in different demes. If the genes initially are in the same deme j migration events require j epochs in different demes and $j+1$ epochs in the same deme, resulting in a shifted geometric distribution with $j-1$ replaced by j, $j = 0, 1 \ldots$, in equation (4.21).

Let $T_S(0,1)$ be the time the two genes spend in the same deme, and $T_D(0,1)$ the time they spend in different demes, given they start in the same deme. Similarly, define $T_S(2,0)$ and $T_D(2,0)$. The two pairs of variables are highly correlated because of the relationship between the number of epochs spent together in one deme and apart in different demes. Obviously, $T(2,0) = T_S(2,0) + T_D(2,0)$ and $T(0,1) = T_S(0,1) + T_D(0,1)$. We find that

$$T_S(2,0) \sim \text{Exp}(d), \qquad (4.22)$$

and

$$T_D(2,0) \sim \mathrm{Exp}\left(\frac{Md}{(M+1)(d-1)}\right). \qquad (4.23)$$

The variables $T_S(2,0)$ and $T_S(0,1)$ have the same distribution, whereas $T_D(2,0)$ and $T_D(0,1)$ have similar but not identical distributions. With probability $1/(M+1)$, $T_S(0,1) = 0$, because the two genes coalesce before one of them migrates and hence they do not spend any time apart in different demes. Conditional that one of them migrates (with probability $M/(M+1)$) $T_D(0,1)$ becomes distributed like $T_D(2,0)$. The moments of $T(2,0) = T_S(2,0) + T_D(2,0)$ follow now from equations (4.22) and (4.23), and

$$\mathrm{Cov}(T_S(2,0), T_D(2,0)) = \frac{d-1}{d^2}. \qquad (4.24)$$

The moments of $T(0,1) = T_S(0,1) + T_D(0,1)$ follow similarly, for example,

$$E(T(0,1)) = E(T_S(2,0)) + \frac{M}{M+1}E(T_D(2,0)), \qquad (4.25)$$

where the relations mentioned above have been applied. The characterisations given in equations (4.22)–(4.24) do not generalise easily to higher sample sizes.

In the coalescent process with population structure and limited migration, a typical outcome of the process is a series of rapid coalescence events within demes followed by long waiting times until the last lineages are in the same deme. This shape of the genealogy would lead to less singletons and more intermediate frequency variants in sequences than expected under the basic coalescent model. Figure 4.11 shows the frequency distribution for ten genes sampled from each of two demes, with $M = 0.1$. Comparing with Figure 4.7, intermediate frequency variants are much more common. Tajima's D would thus be expected to be positive when sampling equal numbers of genes from different demes when migration is limited. Figure 4.12 shows mismatch distributions for six simulated data sets with $M = 0.1$ and ten genes sampled from each of two demes. These distributions are closer to being bimodal than for a panmictic population, Figure 4.5.

Equation (4.17) has the interesting implication that the effective population size for a model with migration rate M and genes initially sampled from the same population is $2Nd$, that is, the approximating basic coalescent has constant size $2Nd$.

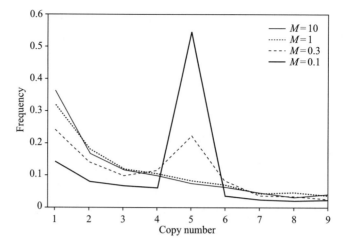

Figure 4.11   The frequency distribution in a sample of ten genes, five from each of $d = 2$ demes. Compare Figure 4.7.

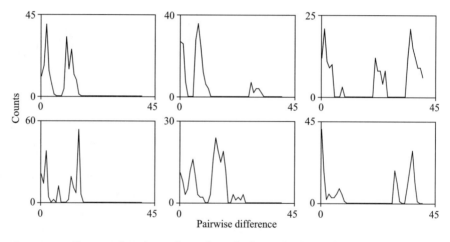

Figure 4.12   Six examples of genealogies from the finite island model with $d = 2$ and $M = 0.1$. The scaled mutation rate is set to $\theta = 3.63$ in order to give the same expected number of segregating sites as in a panmictic population with $\theta = 10$, as was illustrated in Figure 4.5.

4.6.3   General models of subdivision

4.6.3.1   *Stepping-stone models*

Island models assume that a gene from one island or deme is equally likely to migrate to any other island. This assumption is relaxed in stepping-stone models, where the demes typically are arranged in a one- or two-dimensional grid, such that migration only occurs between adjacent demes in the grid, and possibly at different rates. The equally sized demes are replaced by demes of arbitrary sizes, such that $2N_i$ is the size of deme

i, and $M_{ij} = 4N_i m_{ij}$ is the rate of migration from deme i to deme j (see Figure 4.13). Analytical results for stepping-stone models are limited, but simulation of genealogies and sample configurations can be done easily and analogously to simulation in the island model.

4.6.3.2 Continuous space models

Recently, progress has been made in treating space as a continuous entity, with genes diffusing in space until they get sufficiently close for coalescence events to occur. Figure 4.14 shows such diffusions in one and two dimensions. In one dimension, coalescence events occur when diffusing genes cross each other. Wilkins and Wakeley (2002) derived the distribution of possible genealogical histories in one dimension in space of finite length and reflecting borders.

Analysis of the model revealed that (1) the expected coalescence time between two genes increases monotonically with the physical distance between them; (2) the place of the most ancient coalescence events are biased towards the center of the population; and (3) genes that have not coalesced at a given time will be biased towards the edges of the population.

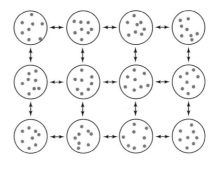

Figure 4.13 Two-dimensional stepping-stone model.

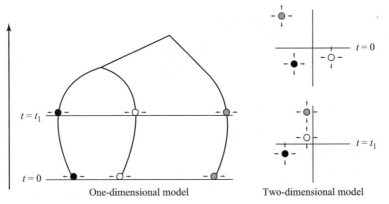

One-dimensional model   Two-dimensional model

Figure 4.14 Continuous space models in one- and two-dimensions. Shown are the positions of genes at two time points; coalescence events occur if genes are sufficiently close to each other.

In two dimensions, the situation is more complex because diffusing genes will never meet at exactly the same spot. Barton et al. (2002) defined a neighbourhood size such that coalescence events occur whenever two genes are within the same neighbourhood (see Figure 4.14).

4.6.4 Non-equilibrium models

Common to all models of population structure discussed so far is that they assume the processes giving rise to population structure have reached a dynamic equilibrium. In particular, under a dynamic equilibrium the genealogy of a sample of genes is independent of when the genes have been sampled. (Coalescent models with fluctuating population sizes are not in dynamic equilibrium.) Dynamic equilibrium is a convenient assumption that facilitates easy ways of simulating sample configurations and genealogies, and in some instances leads to closed form expressions of important quantities. However, many real populations cannot be assumed to be in dynamic equilibrium. This depends largely on the estimated effective population size for the particular real population. If N_e is large one unit of coalescence time might correspond to many thousands of years. For example, for Drosophila N_e is typically around 10^6 which corresponds to about 1 million years (with a generation time of about 1 year). Even in humans, an effective size of 10^4 corresponds roughly to 200,000 years counting one generation as 20 years.

Population sizes and other demographic parameters have often not stayed constant over such a period. In particular, populations split up and evolve independently or populations that have been separated for a long time may merge again. Both of these processes can be modelled using time as an explicit parameter. Tractable cases include what happens over time after a population has split into two, and the opposite situation, what happens over time when two distinct populations merge.

4.6.4.1 *Splitting of populations*

Figure 4.15 shows a situation where a population at some time T in the past splits into two populations that potentially exchange migrants (Nielsen and Wakeley 2001). In this coalescent process time is measured in units of $2N_1$ generations, implying that also T is given in units of $2N_1$ generations. Assume that at a given point in time there are k_1 ancestral genes in population 1 and k_2 ancestral genes in population 2. Looking back in time, coalescence events occur in population 1 and population 2 at intensity $\binom{k_1}{2}$ and $\binom{k_2}{2}N_2/N_1$, respectively, up until the time of population divergence T. Migration events occur at rates $k_1 M_1/2$ and $k_2 M_2/2$, where $M_i = 4N_i m_i$, $i = 1, 2$ is the population scaled migration rate by which genes are leaving population i to enter the other population. At time T, the two populations merge (looking backward in time) and a new ancestral sample is created

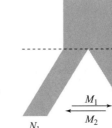

Figure 4.15   Model for splitting of two populations. The parameter M_i is the scaled migration rate of genes leaving population i to join the other population. At time T the two populations merge to form one large ancestral population and hereafter genes are mating randomly.

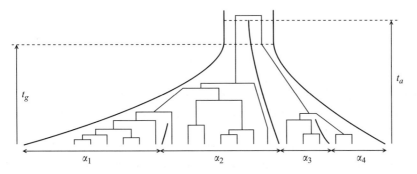

Figure 4.16   Combined model for growth (exponential) and splitting of populations. α_i is the size of subpopulation i, $\sum_{i=1}^{4} \alpha_i$ is the total present day population size, and t_g is the time to population stopped decreasing. From there on a constant population size is assumed. t_a is the deepest split affecting the genealogy of the sample. Splitting events can occur at any time, but no migration is allowed after splitting.

by letting $k_A = k_1 + k_2$ and joining the two samples into one big sample. Thereafter, coalescence events occur at rate $\binom{k_A}{2} N_A/N_1$, while there are k_A ancestral lineages, until there is only one lineage left.

Wilson et al. (2003) proposed a more complex model targeted to the current understanding of the human population. The model is illustrated in Figure 4.16. The model assumes that the population is constant until some time in the past where it starts to grow exponentially. Furthermore, at arbitrary (stochastic) points in time the population splits into subpopulations. However, the model does not allow for migration between subpopulations as in Nielsen and Wakeley's model, and subpopulations do not merge to form larger populations.

It is difficult to predict how the combined effect of exponential growth and population splitting in this model would affect summary statistics such as Tajima's D. For application to human evolution and dispersal obvious further extensions to this model would be to allow for migration between subpopulations, and for recombination (see Chapter 5).

4.7   Coalescent with balancing selection

Balancing selection is selection that leads to the maintenance of a poly-morphism in a population. This can be either through overdominance (heterozygotes are more fit than homozygotes) or through negative, fre-quency dependent selection (a given allele is selected for when it is rare and selected against when it is common). The findings in the 1960s and 1970s of abundant enzyme polymorphisms in electrophoretic studies led many researchers to conclude that balancing selection is prevalent. However, the abundant variation observed would imply a very high genetic load if selec-tion is prevalent, and the apparent presence of the molecular clock together with the first DNA sequencing results were at odds with the selectionist view. This led to the Neutral Theory (Kimura 1968), which states that most of the variation observed is selectively neutral, because deleterious alleles are rapidly removed, and the few advantageous alleles rapidly fixed. The growing number of surveys of DNA sequence variation at the popu-lation level, however, has revived the idea that balancing selection might be quite common. The reason for these claims is that the variation observed is incompatible with the neutral coalescent in that there is a very high level of variation in a limited region of the sequence, and that the sequences can be separated into distinct groups with little variation within a group and a large divergence between groups. Both of these observations are compat-ible with the coalescent with balancing selection outlined below. It should be noted, though, that the observation is also compatible with mixing of formerly isolated populations, which should also be considered before using balancing selection as the favoured explanation for the results.

When it occurs balancing selection is most often between just two alleles at a gene. Well-known examples are Sickle-cell anemia in areas with a high incidence of malaria, and likely the ADH gene of *Drosophila melanogaster* reported in Kreitman's seminal 1983 paper. Multiallelic balancing selec-tion occurs in some special, but important, genetic systems, and will be considered below.

4.7.1   Two allele balancing selection

To illustrate the principle of balancing selection on the coalescent process we use a simple model of one locus with two alleles, A and B. The model is visualised in Figure 4.17. These alleles are both kept in the population by strong, balancing selection. Thus, when $2N$ is large we can assume their frequencies in the total population are constant over time with $P(A) = p$ and $P(B) = 1 - p = q$. Each generation, one allele may mutate to the other at rate u. The resulting coalescent process is analogous to a coalescent model with population structure, where the two alleles could be considered two

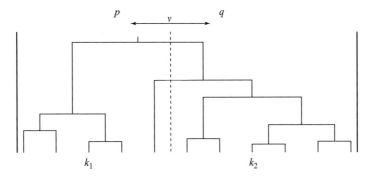

Figure 4.17 Two allele balancing selection divides the population into two types with frequency p and q. Genes are sampled from these types and traced backwards. Movement between the types is determined by the scaled mutation rate v.

different demes, and the scaled mutation rate $v = 4Nu$ between alleles is analogous to the migration rate between the demes.

We shall briefly describe how to construct this coalescent process. Assume a sample of k alleles, with k_1 of type A and k_2 of type B. Two alleles can coalesce only if they are of the same type. In the total population, the population size of genes of type A is $2Np$ and the population size of genes of type B is $2Nq = 2N(1 - p)$. The coalescent intensities for each of the four types of events are given by (with time in units of $2N$ generations):

$$I_{\text{coal } A} = \frac{k_1(k_1 - 1)}{2p}, \tag{4.26}$$

and

$$I_{\text{coal } B} = \frac{k_2(k_2 - 1)}{2q}. \tag{4.27}$$

The mutation intensities are respectively

$$I_{A \rightarrow B} = k_1 \frac{qv}{2p}, \tag{4.28}$$

and

$$I_{B \rightarrow A} = k_2 \frac{pv}{2q}. \tag{4.29}$$

Thus, the time until the first event happens is exponentially distributed with the sum of these four intensities. The process can be simulated by drawing a random number from this distribution to determine the time of the first event, and then subsequently determine the type of the event (coalescent or mutation event) by drawing a random number weighted according to the four intensities above. If a coalescent occurs, the number of samples of one of the two types is decreased by one, if a mutation occurs, one gene has its associated type switched to the other type.

4.7.1.1   *Genealogical consequences of two-allele balancing selection*

The close resemblance of balancing selection with population structure implies that the effect on the shape of the coalescent tree is the same. If the scaled mutation rate ν is small, most genes will coalesce within each type before mutation occurs. This will give a coalescent tree with many short terminal branches corresponding to these rapid coalescences within types and a long branch connecting the two types. Positive values of Tajima's D are thus expected. Exactly the same pattern is expected for the coalescent tree in a subdivided population with small migration rate M. The effects of migration and balancing selection can only be distinguished if more than one locus is analysed, since subdivision will affect all loci equally, whereas balancing selection only affects the loci under selection and some neighbouring loci in tight linkage with the loci under selection.

4.7.2   Multiallelic balancing selection

Examples where multiallelic balancing selection is known to be acting are in incompatibility systems such as plant self-incompatibility, fungal incompatibility, and (most likely) the major histocompatibility system of vertebrates. The genes underlying these systems all contain a very large number of very divergent alleles, making them the most polymorphic genes known. Figure 4.18 shows a phylogeny of self-incompatibility alleles in different species of the night-shade family (Solanaceae). Alleles from a single species, for example, tobacco (*Nicotiana tabacum*) are often more than 50% different at the amino acid level (and at the nucleotide level as well), and it is therefore often observed that pairs of alleles from two different species are much more similar than a pair of alleles from a single species. This phenomenon has been termed trans-specific evolution and indicates that some coalescence events of alleles from a single species predate the divergence of the different species. Thus, if one knows the time of divergence of species (e.g. from fossils), one can estimate the TMRCA in these systems. In Solanaceae, the TMRCA is likely more than 50 million years (Figure 4.18). Such ancient genealogies therefore provide information about much more ancient events than genealogies of most neutral genes (TMRCA on the order of 1 million years or less in humans), and the shape of the genealogies have therefore attracted much attention because they may provide information about ancient events, such as speciation. However, to be able to use this information one needs a coalescent theory for multiallelic balancing selection where only one of each allele type is sampled, thus discounting any

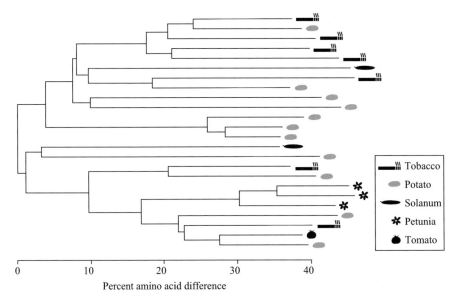

Figure 4.18   The genealogy of self-incompatibility alleles from different species of Solanaceae, potato, tomato, petunia, and tobacco. Adapted from an illustration by Xavier Vekemans (personal communication).

polymorphism that may exist within a given functional allele type. Neutral mutations can then subsequently be added to the genealogy according to Algorithm 3 of Chapter 2.

Takahata (1990) developed such a theory with the major histocompatibility complex in mind, but the theory is easily extendable to self-incompatibility systems as well. We assume symmetrical, balancing selection where all homozygotes have fitness $1 - s$ and all heterozygotes have fitness 1. Some self-incompatibility systems can be modelled by setting $s = 1$ (Vekemans and Slatkin 1994). Because of the strong balancing selection, Takahata assumed that a fixed number of different alleles, k, are being maintained at equal frequencies, $1/k$. This fixed number is the number of common alleles at mutation–selection–drift equilibrium. Sometimes an allele is lost by genetic drift and sometimes a new specificity arises by mutation and invades the population and quickly attains its equilibrium frequency if it is not lost by genetic drift immediately. Takahata approximated this allelic turnover process by assuming that with intervals determined by the average allelic turnover time a random allele is lost and replaced by a descendant of one of the other alleles picked at random. This model is depicted in Figure 4.19 and is an instance of the Moran model discussed in Section 1.11. Since the Moran model is an exchangeable model, the structure of the allele genealogy is therefore the same as the structure of the genealogy in the basic coalescent, but with a different time scale determined by the force of balancing selection. Thus if we measure time in units of

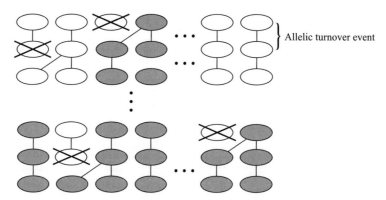

Figure 4.19 Takahata's Moran-model in alleles used as an approximation to strong, multiallelic balancing selection. In each generation one allele is chosen to give birth to a new allele, not seen previously in the population. The new allele replaces a randomly chosen allele. The grey coloured alleles are represented in the current population, white coloured alleles have died out before the present time.

$2 f_s N$ instead of $2N$, where f_s is the elongation of the genealogy caused by selection, the basic coalescent is recovered. Takahata was able to calculate the scaling factor, f_s, as

$$f_s = \frac{\sqrt{s}}{2V} \left\{ \ln \left(\frac{S}{16\pi V^2} \right) \right\}^{-3/2}, \tag{4.30}$$

where $S = 4Ns$ is the scaled selection coefficient and $V = 4Nv$ the rate of invasion of new specificities (the turnover rate). For the human major histocompatibility system estimates on the order of $4Ns = 400 - 4,000$ have been reported, which, depending on assumptions about the mutation rate to new specificities, is equivalent to f_s in the range $10 - 1,000$. Thus, the theory appears to be able to explain the very long retention time of alleles (i.e. the very deep allelic genealogy) in systems with multiallelic balancing selection.

4.7.2.1 *Genealogical consequences of multiallele balancing selection*

Stochastic simulations of multiallelic balancing selection show that Takahata's rather crude approximations are quite accurate when selection is sufficiently strong. Therefore, we do expect that the allelic genealogy has a similar shape to the neutral coalescent model if the mutation rate to new specificities and the population sizes have remained constant. Uyenoyama (1997) derived four measures of 'tree shape' expected under the standard coalescent and compared their distributions (using simulations) with the observed values of these measures derived from actual data from both the MHC and plant incompatibility system. She showed that the genealogies deviate significantly from expectations in that the terminal branches are much longer than expected (or vice versa that the branches close to the root are much shorter than expected). This is quite evident from Figure 4.18. Uyenoyama suggested that this may be due to a slow-down of allelic turnover rates with time due to the accumulation of linked, deleterious variation. Other explanations include that genealogical equilibrium has not yet been reached due to the long retention time of alleles, and that some of the polymorphism therefore may date back almost to the origin of the system. Finally, if the different alleles have recombined during the genealogical history, a star-shaped tree is expected (Schierup et al. 2001).

4.8   Coalescent with directional selection

Our ability to analyse the coalescent with balancing selection rests on the important assumption that selection maintains the frequency of the different alleles constant over time. The two-allele and multiple-allele coalescent approximations to balancing selection cannot be adapted to handle directional selection, because selection will act as a force that changes the frequency of an allele over time. This further implies that the coalescent process cannot be separated from the mutation process, since a given mutation directly influences the number of descendants below the mutation in the coalescent tree. Recently, a clever way around this basic problem was presented by Claudia Neuhauser and Stephen Krone (Neuhauser and Krone 1997), who were able to define a graph, termed the *ancestral selection graph*, which has the coalescent with directional selection embedded as a genealogical tree. The ancestral selection graph can be constructed without knowledge about the mutation process. Subsequently, mutations are added to the graph and the coalescent with directional selection genealogy, which is embedded in the ancestral selection graph, can then be revealed by following a simple set of rules. We shall discuss the case of a model of two alleles, but the framework is sufficiently general to allow more complex models (e.g. modelling weak frequency dependent selection, see Neuhauser 1999).

4.8.1   The ancestral selection graph

The basic idea of the ancestral selection graph is that in addition to coalescent events we also allow branching events back in time. These branching events need to be recorded because they represent potential effects of selection, that is, they represent descendants that a gene would leave if it happens to be of the preferred type (see Figure 4.20). Since we do not know the allelic types of the sample when constructing the graph, we need to record all potential branching events which indicate the possible cases of selection.

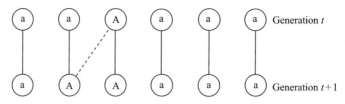

Figure 4.20   Illustration of a hypothetical selection event in a sample of six genes in one generation. In this case gene 2 in generation $t + 1$ has two possible ancestors, gene 2 in the previous generation (the continuing branch), or gene 3 (the incoming branch). Since gene 3 in generation t proves to be of the preferred type, the incoming branch will substitute the continuing branch and the gene 2 in generation $t + 1$ will inherit the preferred type from gene 3 in generation t, see also Table 4.7.

Assume that a given locus has two alleles, A and a, where A has a selective advantage of s, and is hence termed the 'preferred' allele. Thus, the scaled selection coefficient is $\sigma = 4Ns$. Mutation occurs at the same rate u between the two types. If selection is weak, that is, s is on the order of $1/(2N)$, then the ancestral selection graph can be constructed by considering two types of events, (1) the normal coalescent events, and (2) branching events representing hypothetical cases of selection. Given that at a certain time there are j lineages in the population, the intensity of coalescence is

$$I_{coal} = \frac{j(j-1)}{2},\tag{4.31}$$

and the intensity of branching

$$I_{branch} = \frac{\sigma}{2}j.\tag{4.32}$$

If coalescence occurs, j decreases by one, whereas branching increases j by one. When branching occurs, one branch is termed the 'continuing' branch (the branch that would have been there even if there was no selection), and the other branch is termed the 'incoming' branch. This classification is important later when the coalescent tree is to be extracted from the ancestral selection graph.

The process is then repeated with the new value of j until a single ancestor is reached. Since the coalescence intensity is a quadratic expression in j, whereas branching is linearly proportional to j, the coalescent process will ultimately dominate and ensure that a single ancestor (termed the ultimate ancestor) is reached (potentially after many events if σ is large). In the left part of Figure 4.21, a single realisation of the complete process is displayed as the complete ancestral selection graph. Two cases of branching events (putative selection events) have occurred. In each case, the incoming and continuing branches are marked with I and C, respectively.

Once the ancestral selection graph has been constructed, mutations can be superimposed using the Poisson process, as in the case of the basic coalescent. Mutations are added to a branch of length t according to a Poisson distribution with mean $\theta t/2$. Under the simple model studied here, a mutation event will just change the type to the other type. In the middle part of Figure 4.21, mutation events are illustrated with a bullet on the respective branch.

The next step is to add a type to the ultimate ancestor from the stationary distribution at mutation–selection balance. Once the type of the ultimate ancestor is determined, the type of all other nodes in the graph can be determined by tracing the type down through the graph. This is done by changing the type whenever there is a mutation event, using the decision rules in Table 4.7 in cases where two branches meet at a branching event. In Figure 4.21 these rules have been applied to determine the types of all nodes, and then subsequently the true coalescent tree has been extracted (the last part of Figure 4.21). In this case, the ultimate ancestor is also the MRCA, but this is not the case in general.

For moderate levels of selection Krone and Neuhauser (1997) showed that the selection has very little effect on the shape of the genealogy,

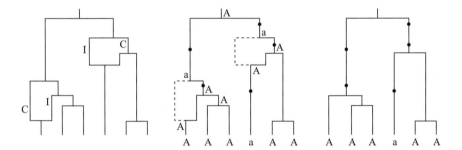

Figure 4.21   The ancestral selection graph for a sample of six genes. In the first graph, branching events are putative selection events. For each of these it is indicated whether they are incoming (I) or continuing (C). In the second graph, mutations have been added (filled circles), the type of the ultimate MRCA added, and the types of the six sampled genes determined by following the rules of Table 4.7. The final tree is the resulting coalescent tree for the case of directional selection.

Table 4.7   Decision table for extracting the coalescent tree from the ancestral selection graph after including mutations

Incoming branch	Continuing branch	Result	Branch kept
a	a	a	Continuing
a	A	A	Continuing
A	a	A	Incoming
A	A	A	Incoming

Table 4.8 Summary table of the effect of different evolutionary forces on the shape of the coalescent tree over the genome

	Whole genome effect	Local effect
Long external branches (Tajima's $D < 0$)	Population growth Very severe bottleneck	Directional selection
Long internal branches (Tajima's $D > 0$)	Population subdivision Less severe bottleneck	Balancing selection Recent population mixing

and therefore weak selection will be difficult to infer from linked neutral variation.

4.9   Summary

The basic coalescent process discussed in Chapter 1 is easily extendable to include various violations of the idealised Wright–Fisher model. This chapter has discussed how demography, population size changes and different forms of selection can be incorporated and how these forces affect the 'shape' of the coalescent tree. This sets expectations for both the sampling variance and the values of summary statistics such as Tajima's D. Table 4.8 summarises the effect of each of the forces discussed on the shape of the coalescent tree.

Recommended reading

Donnelly, P. and Tavaré, S. (1995), 'Coalescents and genealogical structure under neutrality', *Annu. Rev. Genet.* **29**, 401–421.

Hudson, R. R. (1991), 'Gene genealogies and the coalescent process', *Oxford Surveys in Evolutionary Biology* **7**, 1–49.

Nordborg, M. (2000), Coalescent theory, *in* D. J. Balding, M. Bishop, and C. Cannings, eds., 'Handbook in Statistical Genetics', John Wiley and Sons, pp. 179–212.

Slatkin, M. (2000), A coalescent view of population structure, *in* Singh and Krimbas, eds., 'Evolutionary Genetics', Cambridge University Press, pp. 418–429.

5 The coalescent with recombination

5.1 Introduction

Genetic recombination occurs in most organisms on earth, including eucaryotes, bacteria, and viruses. Recombination occurs by quite different mechanisms in these three different types of organisms. In eucaryotes recombination is usually associated with sexual reproduction. Thus, it is generally believed that sexual reproduction exists because it facilitates genetic recombination. No generally accepted theory is available to explain why genetic recombination is so common, but in its favour are arguments that genetic recombination ensures that favourable variants are brought together in the same sequence (or, equivalently, that sequences are created which do not harbour any deleterious variants), and that recombination can maintain more variation which may enable survival over an evolutionary time scale.

The simple coalescence process, including the various extensions in the previous chapter assumes no recombination and this body of theory can therefore strictly speaking only be used on non-recombining sequences. In animals, only Y-chromosomes and perhaps mitochondrion satisfies this constraint, and it is therefore pertinent for practical analysis that recombination can be build into the process. Fortunately, this is possible as shown by Hudson in 1983 very shortly after the coalescent process was first formulated. However, recombination adds much more complexity than any of the extensions discussed in Chapter 4, mainly because no single 'coalescent tree' can describe a sample of recombining sequences. A graph is required to describe the ancestral process which complicates the mathematical description considerably. However, the presence of recombination makes some estimators of evolutionary parameters more accurate than when applied on non-recombining sequences (see Chapter 6). Furthermore, recombination is the key force that enables us to perform linkage disequilibrium (LD) mapping in the search for genes causing common diseases (Chapter 7).

We will begin this chapter with a data example showing some of the hallmarks of recombination. We will discuss how the biology of recombination is captured by simple models of recombination and the alternative form of genetic exchange called gene conversion. We will then formulate the coalescent process with recombination (and its gene conversion counterpart), and finally we will discuss several properties of this complex process that are important for understanding genetic variation in haplotypes and for our ability to detect the presence of recombination.

5.2 Data example with recombination

The Apolipoprotein E locus is an important human gene because variation in the gene is associated with increased risk of Alzheimer's disease. Figure 5.1 shows the segregating sites in a large study by Fullerton et al. (2000), and the different haplotypes, including their frequencies in four different human populations. It can be seen that some haplotypes are much more common than others as expected under the basic coalescent model and that there are groups of similar haplotypes, which are quite different from

73	308	471	545	560	624	832	1163	1522	1575	1998	2440	2907	3106	3673	3937	4036	4075	4951	5229	5361	J	C	N	R	Σ
C	C	A	C	A	T	G	G	G	G	C	G	G	T	T	C	C	C	C	A	G					
															T		T			T	1	0	1	2	4
															T		T			T	1	0	0	1	2
				T	C									T			T			T	0	0	0	1	1
					C										T		T			T	0	0	1	4	5
					C										T		T	C		T	0	0	0	1	1
				T	C	T	C								T						0	0	0	1	1
					C	T	C		T						T						0	0	0	1	1
				T		T	C								T						1	5	2	8	16
						T	C								T						0	1	0	0	1
						T	C								T						8	19	11	7	45
						T	C					G			T						0	0	1	0	1
						T	C	A							T					C	0	0	2	0	2
							C								T						1	0	0	0	1
															T						3	5	0	0	8
	T			T								G			T						1	0	0	0	1
				T											T						2	3	0	0	5
				T							A				T						8	3	3	1	15
					C						A				T						0	0	0	2	2
											A				T						15	6	11	11	43
			T								A				T						1	0	0	0	1
											A				T					C	1	1	4	2	8
								A	A						T					C	0	0	1	0	1
T																		T			2	0	0	0	2
																					1	0	0	1	2
				T														C			0	0	1	1	2
		G		T		T															1	0	0	0	1
		G				T															1	0	0	0	1
					C	T						A									0	0	0	2	2
						T						A					C			C	0	0	1	0	1
						T						A									0	5	9	1	15
						T						A		C							0	0	0	1	1

Figure 5.1 The segregating sites in the study of the Apolipoprotein E gene by Fullerton et al. (2000). Shown are the thirty-one different haplotypes with twenty-one segregating sites compared to the corresponding nucleotides in a chimpanzee sequence. This enables us to see which variant is most likely to be ancestral. The frequencies of each of the thirty-one haplotypes in five human populations are also shown. J = Jackson (African-American), C = Campeche (Hispanic), N = North Karelia (Finnish), and R = Rochester (European-American).

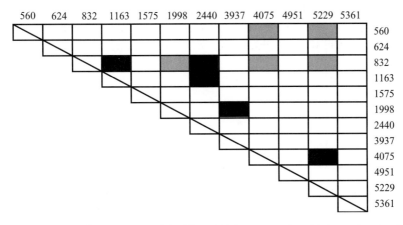

Figure 5.2 Matrix showing cases of significant LD between pairs of informative sites for the Rochester population. Grey hatching denotes significance at the 5% level, black at the 0.1% level. Informative sites are sites where both alleles are present in at least two sequences each.

each other. This can be caused by a deep split in the coalescent tree. However, there are major difficulties in fitting these sequences into a single gene tree as was done in the example of the Y-chromosome in Chapter 1. Many pairs of sites cannot be fitted on a single tree topology without assuming at least three mutations. The pair of sites are therefore said to be incompatible. As an example, for the Rochester population, 32% of all pairs of sites are incompatible. Genetic recombination creates incompatibilities since it allows different parts of the sequence to have topologically different trees. Another sign that recombination has occurred in this data set is the pattern of LD over the sequence. LD is a measure of non-random association of alleles at different sites and is therefore indirectly a measure of the correlation of genealogical trees for different segregating sites. Figure 5.2 shows which pairs of sites are in strong (statistically significant) LD. There is a weak tendency that highly significant LD is found for sites close to each other. Indeed, Figure 5.3 shows that LD is smaller the further apart the segregating sites are. Recombination leads to this pattern since sites far apart experience more recombination events between them than sites close together. Thereby, genealogical trees far apart becomes less correlated than trees close together.

The decrease in correlation with distance is the whole basis for LD mapping, since a site located close to a disease causing variant shares more history with the variant site than sites further away.

A mismatch distribution for the data set is shown in Figure 5.4. The distribution is unimodal which is consistent with exponential growth in the population (see Chapter 4). However, calculation of Tajima's D shows a value close to zero and not a negative value as expected under growth. Thus,

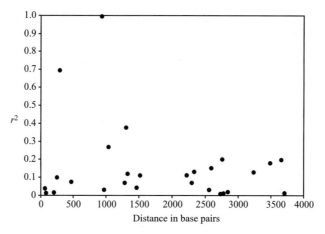

Figure 5.3   LD measured by the r^2 statistic (see Section 7.8 for a definition) using only sites with minor allele frequency 5%, in the Apolipoprotein E data set from the Rochester population.

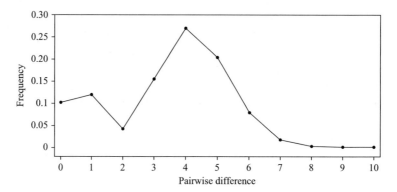

Figure 5.4   The mismatch distribution for the apolipoprotein E data set (see Section 2.5.3). If there is a strong rate of recombination the signature of a tree disappears and all pairs tend to have similar numbers of mismatches. So one should expect a slimmer distribution of mismatches around the same mean in the presence of recombination.

in this case it is more likely that recombination has caused the unimodal mismatch distribution because recombination shuffles variation between sequences, making them more equidistant.

5.3   Modelling recombination

5.3.1   Hudson's model of recombination

Hudson introduced recombination into the coalescent process and presented a simple model in 1983. Even though the model is very simple and does not capture intricate details of the biology of recombination, it still forms the basis for most applications of coalescent theory to recombining sequences. It is very easy to simulate sequence data sets under the model. However, it is much more difficult to conduct inference under this model than under the coalescent models outlined in the previous chapters. This is because the structure needed to describe the relationship of a set of sequences is now a complicated graph, rather than a single tree. Even simple quantities concerning this graph are complicated or impossible to calculate analytically.

Hudson's model of recombination is illustrated in Figure 5.5. Choose a random point uniformly along the chromosome, then copy the genetic material from one parent chromosome to the left of this point and copy the genetic material from the other parent chromosome to the right of this point. The coalescent model has a time reversed perspective on this. Recombination splits the genetic material of a sampled sequence onto two different ancestors such that each position has exactly one ancestor. In this respect recombination events are the opposite of coalescent events that combine two sampled sequences into one ancestor. Hudson (1983a) formulated the coalescent process with recombination as competing exponentially distributed, and independent, waiting times for coalescent and recombination events. The parameter of the exponential distribution determining the coalescent intensity depends on the number of ancestors carrying ancestral material to the sample, whereas the parameter of the exponential distribution for the intensity of recombination depends on the recombination

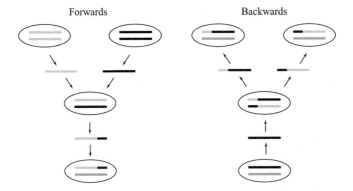

Figure 5.5 Hudson's recombination model on a continuous representation of a sequence. A sequence is made by recombination when an individual creates a haploid genome (sperm cell or egg). Looking forwards, two sequences are recombined into one recombinant sequence. Knowing the allelic states of the grandparent's chromosomes determines one of the child's two chromosomes; the other, the dark grey, originates from the second set of grandparents. Looking backwards, an individual chooses a chromosome from a parent. This chromosome is split onto two grandparental chromosomes. The child's dark grey chromosome is inherited through the other parent and the dark grey chromosomes in grandparents have unknown allelic states.

rate over the sequence, usually termed ρ, times the number of ancestral lineages. Once these intensities have been specified, it is easy to devise an algorithm for simulating the process. A single realisation of the process includes splitting (recombination) and joining (coalescence) of ancestors until a single ancestor is produced, termed the *grand-most recent common ancestor* (GMRCA). The resulting structure is a graph rather than a tree, with coalescent trees embedded, one tree for each position on the sequence.

5.3.2 Biological features of recombination

Hudson's model of recombination simplifies the biological facts about the recombination process. Unfortunately, the recombination process is still not very well understood at the molecular level. Thus, it is only possible to point to some potential deviations from the model that may be of importance in a more sophisticated model. This has already been done to some extent, but is expected to be an active field of research in the future as more data accumulates and the biological understanding of the recombination process increases further.

Two different types of genetic exchange may occur between homologous pieces of DNA, namely what here is called recombination and gene conversion. Homologous gene conversion is the substitution of a small fragment of DNA from one chromosome to another. In a modelling perspective it can be thought of as two very close crossover points even though this rarely is the mechanical way it occurs. Gene conversion is important mainly at the very local level, that is, at the gene level. The mechanisms of recombination and gene conversion are very different in eucaryotes, bacteria, and viruses. For a more in-depth description of the processes, consult Lewin (2003).

5.3.2.1 *Viral and bacterial recombination*

In bacteria, recombination occurs by transfer of part of the haploid genome from one cell to a different cell by invasion of a naked DNA string, transfection or transduction. If single-stranded DNA enters the cell it can intrude into the homologous part of the unwound double-stranded DNA of the recipient. Heteroduplex DNA is then formed and can extend the length of the intruding DNA. Heteroduplex DNA contains mismatches in the double-stranded DNA corresponding to the differences between the recipient and donor strings. Repair mechanisms involving the well characterised Rec protein family will then correct the heteroduplex. If the donor string is used as the template, the affected part of donor DNA is effectively recombined.

In viruses, recombination most often occurs through template switching during the replication process. The genetic material of the virus is replicated by either a DNA- or an RNA-polymerase depending on the type of virus. This polymerase may then jump from one possible template to a different one. This can occur if the two templates have come into the same host cell and in close proximity within the cell. Thus recombination rates in viruses are believed to be determined by the rate of co-infection of the same cell by different variants of the same virus. Because of this, recombination often occurs at several points throughout the genome when different templates get close. This phenomenon is called negative interference, that is, the probability of recombination occurring in a specific point increases if recombination is occurring nearby. Explicit modelling of this in a coalescent framework has not been done.

5.3.2.2 *Eucaryotic recombination*

In eucaryotes, accidental breakage of a single strand in one of the two homologous chromosomes can lead to invasion of the broken strand into the other chromosome, again leading to a region of heteroduplex DNA (see Figure 5.6). Little is known about the average size of the heteroduplex region, which is likely to be species dependent, but it may extend several kilobases. The mismatches in the heteroduplex is subsequently recognised by the DNA repair system which will use one or the other strand (or perhaps both) as templates for repair.

True recombination only occurs if both strands break simultaneously or if the flanking chromosomal parts are exchanged after strand invasion has occurred (see Figure 5.7). If either happens, a Holliday junction is formed which will be visible in the microscope. Many proteins take part in the maintenance and final resolution of the Holliday junction (together forming the so-called 'recombination nodule'). Figure 5.7 shows the formation and resolution of a Holliday junction. The upper left Figure 5.7 shows two double helixes physically aligned; the upper right is the Holliday structure, where two single DNA strands in the DNA double helix have been broken and rejoined with their homologous partner in the other double helix. Due to the high similarity between the homologous chromosome (in humans, identity is more than 99.9%), the Holliday structure can quite freely migrate left or right. Heteroduplexes will then be created whenever there are differences between the homologous chromosomes, because the nucleotides then cannot form Watson–Crick base pairs (C with G and T with A). Depending on the resolution (recreation of two independent DNA

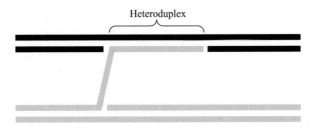

Figure 5.6 Invasion of a single strand into the double-stranded DNA of the homologous chromosome.

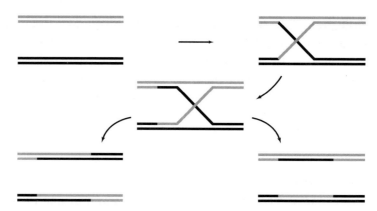

Figure 5.7   Recombination, gene conversion and the Holliday structure. In the first case a gene conversion with exchange of flanking region happens; in the second a pure gene conversion happens. The former is called a recombination event, the latter a gene conversion event. Only one of the two double-stranded DNAs is transmitted to the offspring.

double helixes) of the Holliday structure, the creation of heteroduplex, the potential mismatch repair mechanism and the timing of the subsequent DNA replication, different daughter molecules will be created. What is of interest is the ancestry of individual nucleotides in the daughter molecules with respect to the parent nucleotides. Resolution of the Holliday structure after strand migration can occur either as a gene conversion event with crossing over (Figure 5.7 lower left) or as a gene conversion event without crossing over. Henceforth, we will reserve the phrase 'recombination event' for the former type of event, and the phrase 'gene conversion event' for the latter.

Holliday's model is simplistic and more realistic models have been proposed (see Lewin 2003, Kauppi et al. 2004). However, Holliday's model basically captures how recombination and gene conversion are believed to take place. Other models often lack the symmetry of the strand exchange shown in Figure 5.7.

5.3.2.3   *Relative rates of gene conversion versus recombination in eucaryotes*

The importance of gene conversion versus recombination in eucaryotes depends on the relative rates of single strand versus double strand breaks and the repair mechanisms. The relative frequency of gene conversion to recombination is dependent on how the Holliday junction is resolved. A random choice of cutting two of the four single strands that would create two homologous DNA sequences would lead to equal frequency of the two. In reality gene conversion can be much more frequent than recombination.

Gene conversion has been demonstrated to occur at quite high rates in both humans and Drosophila, both by direct means and by indirect means (analysis of sequences, for example, Andalfatto and Nordborg (1998) in Drosophila, and Frisse et al. (2001) in humans). Crude estimates set the rate of gene conversion to recombination in the range 2–10, with gene conversion relatively more important in regions of the genome with low overall rates of recombination.

5.3.2.4 The effects of alternative repair mechanisms of the heteroduplex

In Figure 5.8 one of the two double helices that emerge from the resolution of the Holliday junction is shown (cf. Figure 5.7). If the two double helices both duplicate, four different double helices could be created (if there is a mismatch in the heteroduplex). The ancestor of a dinucleotide in one of the four double helices would be determined by the 'colour' of the single strand used to create that double helix. However, normally a repair mechanism corrects mismatches in the heteroduplex region. This might happen in different ways, or according to different repair rules. Figure 5.8 shows three alternative ways a heteroduplex region can be repaired. One sequence can be chosen as master sequence and the other repaired according to the master sequence. (1) The top strand is the master sequence and the result is identical to one of the original double helices, that is, there is only one ancestor of the sequence. (2) The bottom strand is the master sequence and the result is a mix of the two original double helices: The flanking regions share an ancestor, the middle region has another ancestor. (3) Alternatively, a master dinucleotide could be chosen randomly for each dinucleotide such that each dinucleotide is either derived from parent 1 or parent 2.

It is important to understand that the repair rule essentially determines which of the original two double helices carry the ancestral dinucleotide. Assume the heteroduplex is repaired according to a master sequence,

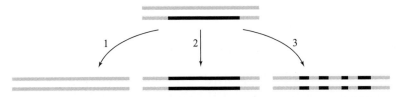

Figure 5.8 Ancestry and the repair mechanism. At the top two sets of double helices have swapped a single strand, which can create mismatching base pairs. Only one of the two resulting double helices is shown (cf. Figure 5.7). Three ways of correcting mismatches are shown: (1) The top sequence is the master sequence, the bottom is corrected according to the top sequence, (2) the bottom sequence is the master sequence, and (3) the mismatches are resolved independently one by one. It gives a mosaic ancestry where different nucleotides originate from different parent strands.

for example,

Before repair   After repair
A A C G A T   A A C G A T
T T T C T A   T T G C T A

such that the third position is a mismatch. If the heteroduplex is repaired according to the top sequence, the original double helix with the top sequence is the ancestor of all six dinucleotides, despite, for example, that in the first position, A physically originates from one double helix, T from another. If the repair rule is to choose randomly with probability $1/2$ then the resolved heteroduplex occurs with probability $1/2$, but the top sequence is no longer likely to be the ancestor of all six dinucleotides. This happens with probability $1/2^6 = 1/64 \approx 0.016$ only.

This very schematic presentation poses questions which need to be answered if the recombination process including gene conversion should be modelled in a realistic way.

1. What is the nature of the strand migration? Is it unidirectional or bidirectional. Is it a random walk or is it energetically driven? The answer to these questions will determine the distribution of the length of the tract of converted material. If it was unidirectional with each step having the same probability, the geometric/exponential distribution would be natural. If it was bidirectional under the same circumstances and not biased against heteroduplex, it would be the distribution obtained by a random walk. Were the random walk to be stopped after a fixed number of steps the suitable distributions would be the binomial/normal distributions. More complicated distributions would be necessary if it was stopped after a random number of steps. If there was a bias against heteroduplex, the length of gene conversions would be sequences dependent, which is computationally difficult as it creates a dependence on the mutational process.

2. Does single stranded invasion occur independently of where along the double helix it is initiated and independently of what the allelic state is of the region where it takes places?

3. How is the heteroduplex resolved? Correcting mismatching nucleotides by randomly changing one of the nucleotides will look very different from not correcting it and just letting the DNA double strand duplicate or using one of the strands as the master sequence (Figure 5.8). Note that the examples of repair rules given here are independent of the allelic state of dinucleotides. There are indications that this might not always be true, that is, that some mismatch pairs are more likely to be changed into certain dinucleotides than others. For example, the mismatch pair AG might be changed to CG more often than to AT.

Until more data has accumulated and we can get firm answers to these questions, Hudson's simple model of recombination and Wiuf and Hein's (2000) model of gene conversion are convenient approximations. Both are discussed in detail below. Wiuf and Hein's model of gene conversion assumes that gene conversion occurs with a rate γ such that the relative impact of gene conversion to recombination is given by the ratio γ/ρ. Correction of mismatches in the heteroduplex occur with a randomly chosen master sequence. Empirical studies suggest that $\gamma/\rho \in (2, 10)$ for humans and Drosophila. Evidence from Drosophila points to a geometric distribution of the gene conversion tract length with mean approximately 350 bp (Marais 2003). In other organisms, the average length is estimated to span from about 100 bp (mice) to 1 kb (yeast), Marais (2003).

Despite the poverty of empirical evidence, this simplistic model of recombination and gene conversion serves as a useful first approximation. Further, it has the advantage that the recombination/gene conversion process is disentangled from the mutational process and thus that the genealogy of a sample can be built first, then mutations can be imposed onto the genealogy, similarly to what was done for the coalescent without recombination as discussed in Chapter 2.

5.4 The Wright–Fisher model with recombination

In order to derive the coalescent with recombination we first return to the Wright–Fisher model of the previous chapter. This model can be modified to include recombination. Each individual creates a haploid genome by recombination each time they contribute to the next generation. An individual in the next generation is then made by choosing one haploid genome

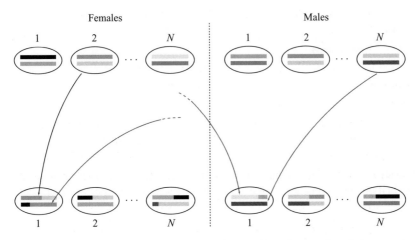

Figure 5.9 The diploid Wright–Fisher model with recombination.

from the male set and one haploid from the female set. This process is illustrated in Figure 5.9. We consider a small region (sequence) within the genome. A recombination event takes place in the region with probability r, in which case a point is chosen uniformly along the paternal and maternal sequences and they recombine in that point. The existence of the gene conversion is ignored.

If this is viewed backwards in time, the effect of recombination will be that the ancestral material to a specific DNA sequences is found on two DNA sequences in the parent, which again came from two different grandparents, etc. In the generation before a sequence was created by recombination, there would have been one more sequence carrying ancestral material than after.

If we focus on a single point on the sequence, it will be inherited from one parent only, thus the Wright–Fisher model with recombination reduces to the Wright–Fisher model without recombination for each point on the sequence, but different points on the sequence are correlated instances of the Wright–Fisher process without recombination. The tree relating the sequences in a single position is called the *local tree* of that position. Thus, the genealogy of the whole sequence can be seen as a collection of local trees, one for each position.

In analogy with the Wright–Fisher model without recombination, we may ignore the existence of individuals and describe the process as acting only on individual sequences as illustrated in Figure 5.10. When the number of sampled sequences is small relative to the population size, and inbreeding is limited the quantitative effects of ignoring which sequences are present in the same individual are negligible. The only difference to the genuine haploid model is that each sequence must choose two parents and that a sequence is created after letting the parent sequences recombine (if they do). If the recombination rate is low, there is a high probability of no recombination and the new sequence will be a direct copy of one of the two parent sequences.

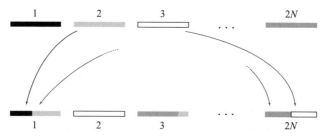

Figure 5.10   The haploid Wright–Fisher model with recombination. Each haploid individual chooses two parents that make the individual's chromosome by recombination. Real biological individuals can be made by packing pairs of haploid individuals together. Describing the process as a purely chromosome focused process has negligible consequence for the actual probabilities involved. Recombination happens with probability r.

Figure 5.11   In a large population a sequence created by recombination is always made up of two sequences that have no other descendants in the sample. Thus, the left part of the figure is possible, whereas the right part is impossible because one ancestral sequence is involved in both a recombination event and a coalescence event at the same time.

5.5   Algorithms

Recombination and coalescent are competing processes that determine the graph structure of the genealogy of the sample, with recombination causing splitting and coalescent causing merging of sequences when looked at backwards in time. Analogous to the definition of the scaled mutation rate θ, the scaled recombination rate is defined as $\rho = 4Nr$. It is convenient to represent a sequence by a continuous interval (infinite sites model) of length $\rho/2 = 2Nr$, i.e. the expected number of recombination events in the population in one generation. ρ is called the scaled recombination rate or the population recombination rate. There are two alternative algorithms to construct the coalescent with recombination process, Hudson's back-in-time algorithm and Wiuf and Hein's (1999b) spatial algorithm. They illustrate different aspects of the process. Hudson's algorithm is simplest and most useful in the majority of applications.

5.5.1   The ancestral recombination graph

Assume the history of n sampled sequences is being described going backwards in time and that the first event encountered is a recombination event. How will that affect the number of ancestors to the sample and the distribution of genetic material ancestral to the sample? Before the recombination event (i.e. closer to the present), there were n sequences each carrying the ancestral material to the n sequences in the sample. After the recombination event (further back in time), one of the sequences had two ancestor sequences—one carrying ancestral material to the left of the recombination break point and one carrying ancestral material to the right of the recombination point.

If time is measured discretely in generation, the time until a recombination event occurs is geometrically distributed with parameter $r = \rho/4N$. Tracing a sequence back in time, the probability that it was created by recombination j generations back is

$$P(T_R = j) = r(1-r)^{j-1}, \tag{5.1}$$

where T_R denotes the number of generations until the first recombination event. For the continuous time approximation, the waiting time until a recombination event occurs in a given sequence is exponentially distributed with parameter $\rho/2$. It follows by rescaling time,

$$P(T_R^c \le t) = 1 - (1 - r)^j = 1 - \left(1 - \frac{2Nr}{2N}\right)^{2Nt} \approx 1 - e^{-\rho t/2}, \quad (5.2)$$

where $j = 2Nt$. If there are currently k sequences ancestral to the sample then the time to the next recombination event is exponentially distributed with parameter $k\rho/2$. When a recombination event occurs, it is equally likely to occur in any of the k ancestors and the position of the recombination breakpoint in the chosen sequence is picked uniformly over the sequence length.

If $2N$ is large it is unlikely that a sequence is involved in both a recombination event and a coalescence event at the same time (Figure 5.11). In discrete time the probability of a coalescence is $1/(2N)$ and the probability of a recombination in one generation is $r = \rho/(4N)$ such that the probability the ancestor of two sequences has experienced recombination is $\rho/(8N^2)$, which is negligible for large $2N$ and fixed ρ.

Assuming k sequences, the time to a coalescence event is exponentially distributed with parameter $k(k - 1)/2$ and the time to a recombination event is exponentially distributed with parameter $k\rho/2$ and these two distributions are independent. To devise an algorithm for simulation of the process we first find the time until either a coalescence or recombination event occurs. Since the two exponential distributions are independent this time is exponentially distributed with parameter

$$\frac{k(k - 1)}{2} + \frac{\rho}{2}k, \quad (5.3)$$

and the probability that the first event is a coalescence event is

$$\frac{k - 1}{k - 1 + \rho}. \quad (5.4)$$

The probability that it is a recombination is

$$\frac{\rho}{k - 1 + \rho}. \quad (5.5)$$

If recombination occurs, the number of ancestral sequences is increased by one; if coalescence occurs the number of sequences is decreased by one.

Thus, in the simplest form, the algorithm for simulating a genealogy of a sample of n genes from the coalescent with recombination is:

Algorithm 5

1. Start with $k = n$ genes.

2. For k sequences with ancestral material, draw a random number from the exponential distribution with parameter $k(k-1)/2 + k\rho/2$. This is the time to the next event.

3. With probability $(k-1)/(k-1+\rho)$ the event is a coalescence event, otherwise it is a recombination event.

4. If it is recombination, draw a random sequence and a random point on the sequence. Create an ancestor sequence with the ancestral material to the left of the chosen point and a second ancestor with the ancestral material to the right of the recombination point. Increase the number of ancestral sequences k by one and go to 1.

5. If it is a coalescence event choose two sequences among ancestral sequences at random and merge them into one sequence inheriting the ancestral material to both of the sequences. Decrease k by one. If $k = 1$ end the process, otherwise go to 1.

One might speculate whether a single ancestor is ever reached since recombination may keep distributing genetic material onto more and more ancestors. However, since the coalescent intensity is proportional to the square of the number of ancestors while the recombination intensity is linear in the number of ancestors, a GMRCA (see page 132) is always found, even though it may take a long time. Mathematically, this description of the coalescent with recombination is a birth–death process with birth rate $\binom{k}{2}$ and death rate $k\rho/2$ while there are k lineages. The graph structure resulting from applying the algorithm has been termed the *ancestral recombination graph* (ARG) by Griffiths and Marjoram (1996). This structure includes all information about the history of the sample. For a given position (nucleotide) of the sequence, the local tree can be extracted from the graph. To extract the tree for a given position x trace the point backwards from the present. Each time a recombination event occurs choose the lineage that is ancestral to x. Thus, the graph emphasises that different parts of the sequence have different coalescent trees, perhaps with different times to the MRCA. Actually, each single part of the sequence may have found a MRCA long before the occurrence of the GMRCA.

5.5.1.1 *An example*

An example outcome of the above algorithm for two sequences is illustrated in Figure 5.12. At the present time, that is, the bottom of the graph, two sequences of lengths $\rho/2$ have been sampled. Waiting backwards in time,

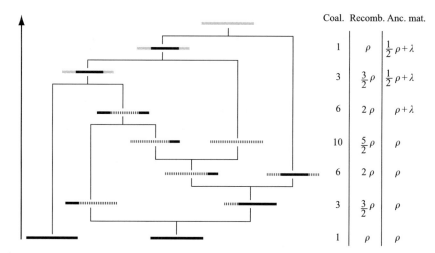

	Coal.	Recomb.	Anc. mat.
1	ρ	$\frac{1}{2}\rho + \lambda$
3	$\frac{3}{2}\rho$	$\frac{1}{2}\rho + \lambda$
6	2ρ	$\rho + \lambda$
10	$\frac{5}{2}\rho$	ρ
6	2ρ	ρ
3	$\frac{3}{2}\rho$	ρ
1	ρ	ρ

Figure 5.12   Black lines represent the sample sequences or ancestral sequence material to these. Dotted lines represent non-ancestral material. Light grey lines indicate that a MRCA has been found. The non-ancestral piece formed after the first coalescence event between two non-consecutive pieces of ancestral material is trapped material. Also shown is the rate of coalescence and recombination, and the amount of material spanned by ancestral material. λ is the length of the black bar in the sequence with dashed ends.

either a coalescent event (with rate $k(k-1)/2 = 1$) or a recombination event (with rate $k\rho/2 = \rho$) could occur. In this example the first event is a recombination event. After the event there are three sequences with ancestral material to the two sampled sequences. The next two events are also recombination events. In one of the two events a sequence is created with no material ancestral to the sample. The rate of a coalescence is now 10, while the rate of recombination is 2.5ρ.

The fourth event is the first coalescent event that also traps a piece of non-ancestral material between two pieces of ancestral material. As long as the flanking regions are linked, their genealogical histories are identical, so if one segment coalesces into another sequence, so does the other. After three more coalescence events all the ancestral material from the two sampled sequences have found common ancestry, in fact have found a GMRCA. There are two MRCAs: one is also the GMRCA which is the MRCA of the middle island of ancestral material, the other is the MRCA of the two flanking islands of ancestral material. This MRCA is created at the second coalescent event. When two pieces of ancestral material are bridged together they share fate as long as they are not cut by recombination again. The material between the two pieces is called *trapped material*.

5.5.1.2   *Discrete versus continuous sequences*
Real sequences have a discrete number of base pairs rather than an infinite number of sites. The infinite sites model described in the previous section can be converted to a discrete model by dividing the continuous interval

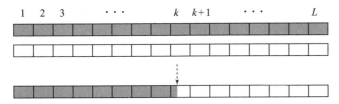

Figure 5.13   Hudson's recombination model in a continuous representation of a sequence with the discrete structure superimposed. The probability of a recombination within a specified dinucleotide is small—about 10^{-8} in humans per generation. In the continuous representation a sequence 1000 bp long in a population with effective population size of 10,000, will then be $\rho/2 = 2 \cdot 10{,}000 \cdot 1000 \cdot 10^{-8} = 0.2$ long. A recombination event within the kth nucleotide counts as a break between the kth and the $(k+1)$th nucleotide in the discrete model.

into equally sized fragments corresponding to nucleotides. The difference between the models is illustrated in Figure 5.13. Thereby there is a constant scaled rate of recombination ρ/L per nucleotide where L is the number of nucleotides. For small sequences of length L, r is approximately Lr_0, where r_0 is the probability of a recombination per generation per nucleotide. We will continue to use the infinite sites model if not stated otherwise.

5.5.1.3   *Improvements of the basic algorithm*

The ARG just described is simple but in many cases unnecessarily time consuming to simulate. In particular when the recombination rate is relatively high. There are two reasons for this:

1. Any given point on the sequence may have reached a MRCA long before the GMRCA.
2. The recombination break point may create ancestors that do not carry material ancestral to the sample.

It is straightforward to make the algorithm more efficient by adjusting for either of these factors. The first factor can be taken care of if a record of the number of ancestors of each position is kept. As soon as a position has found a MRCA it is no longer counted as ancestral material, but as non-ancestral. This reduces the time complexity of the algorithm considerably.

The second factor can be dealt with in several ways. The simplest way is to keep the recombination rate at $k\rho/2$. However, once a recombination occurs, it is recorded whether both recombining sequences carry material ancestral to the sample or not. If only one does, the other can be discarded and the number of ancestral sequences is unaffected by the recombination event. As a consequence, the recombination event does not affect the genealogy at all. An even more efficient way is to make sure that only recombination events that change the distribution of ancestral material are feasible. To do so one keeps track of the ancestral material present on all sequences. A recombination event only affects the genealogy if it occurs in the region spanned by the left and right endpoints of ancestral material on the recombined sequence. This region might include segments of non-ancestral trapped material as well (see Figure 5.12). The intensity, A, of a recombination event is $0 < A \leq k\rho/2$, where A is the sum of material spanned by left and right endpoints of ancestral material.

For recombination rates $\rho > 10$ much efficiency is gained by taking both factors into account, but the efficiency is not greatly increased by keeping track of A as compared to just keeping track of whether all ancestors carry ancestral material or not. An ARG trimmed according to the two factors (and keeping track of A) is called Hudson's algorithm or Hudson's graph. It is not possible to reduce the algorithm further.

5.5.2   Sampling ARGs: Not back in time, but along sequences

The coalescent process has been described as a process starting with n leaves and then coalescing pairs of sequences until a MRCA sequence has been reached. There is an alternative algorithm that moves along the sequences and modifies the genealogy as recombination break points are encountered (Wiuf and Hein 1999*b*). The basic idea is the following: (1) simulate the genealogy for the first position in the sequences; (2) find the first break point as one moves towards the right end of the sequences; (3) choose the sequence that undergoes recombination and modify the genealogy accordingly. Intuitively it is clear that there is a formulation of the coalescent with recombination fulfilling (1)–(3) because we could run through all local trees starting at the left endpoint moving rightwards. However, it is less obvious how the algorithm probabilistically should be formulated. The details are given in Wiuf and Hein (1999*b*); here we will only elaborate on the intuitive formulation of the algorithm.

It works by building up a graph stepwise, a graph that is embedded in the ARG and that contains Hudson's graph. The graph for a given position t in the sequences is called the *local graph* for t and it has the local tree for t embedded. The local graph of position zero is always the local tree but for all other positions the local graph might differ from the local tree. The local graph for t is described relative to the starting point of the sequences. Thus if the starting point is moved (e.g. ignoring a part of the sequences) the local graph for t is given relative to the new starting point.

The algorithm is illustrated in Figure 5.14 with three sequences. A genealogy is sampled for three sequences that describes the relationship of the sequences in the left-most point, zero. This is the local graph in position zero which is an ordinary coalescent tree. Now we would like to move along the sequences scanning for the first break point, that is, the first point

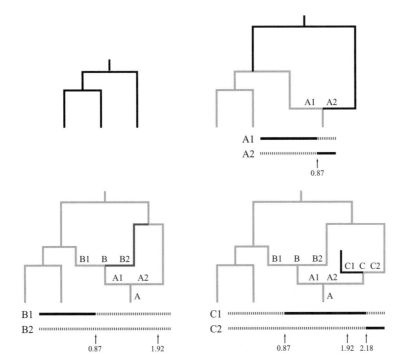

Figure 5.14   The spatial algorithm. The steps are explained in detail in the text. Events that do not contribute to the genealogy of the sequences might be included in the graph created by the spatial algorithm. A graph with three events A, B, and C, is shown creating sequences A1, A2, B1, B2, C1, and C2, respectively. The second recombination event creates an 'empty' sequence: If C2 created after the third event merges with B2 then event B becomes part of the genealogy, whereas if C2 merges with anything else B is not part of the genealogy.

described by a different genealogy than the one describing position zero. Where this point is, depends on the total length of the local graph (tree) in position zero a large graph spans many more generations than a small graph (tree) and the first break point would be closer to 0 in a large graph than in a small graph (Figure 5.15).

In the example, assume the total branch length is 1.8. A variable is taken from an exponential distribution with 1.8 as intensity parameter—here the outcome of the variable is 0.87. Now choose a uniformly random point on the first local tree and postulate a recombination at that point. All positions from 0 to 0.87 ($0 \leq t \leq 0.87$) share the same local tree or local graph; from position 0.87 ($0.87 < t$) the graphs are different. The newly created sequence coalesces with the local graph for position 0.87. In the example the sequence coalesces with the root sequence and the local graph in position 0.87 is not a tree but a graph. Assume the total length of all branches in the local graph for 0.87 is 3.3. The steps are now repeated. An exponential variable with intensity 3.3 is drawn—here 1.05—and adding it to 0.87 to find the location of the next break point. Now choose a recombination event

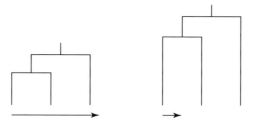

Figure 5.15   Two genealogies relating the first position in three sequences. The second tree has higher total branch length than the first. In each generation there is a small chance of a recombination event between any two nucleotides and in consequence the first recombination break point moving rightwards tends to be closer to the origin in the second tree than in the first tree. Under the infinite sites assumption the length (in units of scaled recombination distance) until the first break point is exponentially distributed with intensity the total branch length.

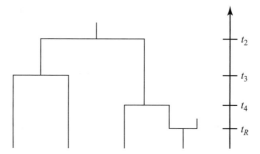

Figure 5.16   Build a coalescent tree by adding sequences. The sequence created by recombination at time t_R coalesces with the rest of the genealogy at rate 4 while in the interval (t_R, t_4), at rate 3 in (t_4, t_3), at rate 2 in (t_3, t_2), and at rate 1 after time t_2. Generally, it is not an advantage to build the tree in this way because it requires more book-keeping and simulation of more waiting times than if the tree is built in the standard 'back in time' way.

uniformly on the branches with total length 3.3, create a new sequence by recombination and let it coalesce to the branches of the local graph for $0.87 + 1.05 = 1.92$ (which is the same as the local graphs in $0.87 < t \leq 1.92$). This is continued until the domain of the local graph has covered the sequences and a complete history has been created.

This algorithm uses the property that the coalescent also can be made by adding single sequences to a growing tree as illustrated in Figure 5.16. If a genealogy has been sampled for k sequences, it can be turned into a genealogy for $(k + 1)$ sequences by taking the $(k + 1)$th sequence in the present and letting it coalesce to the genealogy of the k sequences. The sequence coalesces at rate k to the k branches in the epoch with k branches, at rate $(k - 1)$ in the epoch with $(k - 1)$ branches, etc. The rate for i $(i = 1, 2, \ldots, k)$ follows from the fact that the sequence forms i pairs with the other sequences and each pair coalesces at rate 1.

The spatial algorithm is slower than Hudson's algorithm, but faster than the ARG. Note that the length of the local graph increases as recombination break points are encountered. Thus, recombination break points become denser and denser positioned as one moves along the sequences. (It is possible to reduce the number of recombinations kept track of. We will not pursue this here.) Most of these break points are not part of the genealogy of the sequences, that is they occur in non-ancestral material or material that have already found a MRCA. In general, it cannot be known in advance whether these extra break points affect the genealogy or not; it is first known after completion of the graph (Figure 5.14).

However, this algorithm underlines two interesting properties of the coalescent with recombination process as illustrated in Figure 5.15. First, the process is not Markovian in the sense that it is not enough to know the local tree to assign the probability of the next local tree. Earlier local trees must be known as well, since a new sequence created by recombination can also coalesce to branches of earlier local trees. Second, local trees cannot be reduced to coalescent topologies or unrooted tree topologies, because trees with long branches will on average encounter a recombination break point sooner than trees with short branches. Methods that try to reconstruct the sequence history by using HMMs (Hidden Markov Models) with unrooted tree topologies as the hidden states, violate both these properties.

Additionally, this algorithms underlines the effect of a recombination as a subtree transfer between neighbouring genealogies (see Section 5.11.1).

5.5.3   Efficiency of different algorithms

Hudson and Kaplan (1985) calculated the expected number, R_n, of recombination events in a sample's history

$$E(R_n) = \rho \sum_{i=1}^{n-1} \frac{1}{i}, \qquad (5.6)$$

which is proportional to the total length of the genealogy of a single site. The number R_n only counts events that fall within ancestral material— not events that fall in trapped material, outside ancestral material, or in material that have already found a MRCA. The expected number, R_n^*, of events before the GMRCA can also be found. The exact expression is of little interest, what matters here is how the number depends on the sample size n and the rate ρ. Ethier and Griffiths (1990) showed that

$$E(R_n^*) \approx \rho \log(n) + e^\rho, \qquad (5.7)$$

which is true at least for large n and/or ρ. Now the numbers in equation (5.6) and (5.7) can be quite different, for example, $E(R_n^*)$ increases at an exponential rate with ρ, whereas $E(R_n)$ is linear in ρ. It is thus natural to expect that Hudson's algorithm is advantageous computationally. This is indeed the case: Let E be the total number of events in the graph (irrespective of what algorithm is used), then

$$E = 2R + n - 1, \qquad (5.8)$$

where R denotes the number of recombination events. This number depends on the chosen algorithm (and is a stochastic variable in itself). Each recombination event creates a new sequence and an extra coalescent event is required to complete the genealogy, thereby adding two events to E. The number $R_n + R_n^T$, where R_n^T is the number of recombination events in trapped material, is always a lower bound to R because all events in trapped and ancestral material modify the genealogy. Thus

$$2R_n + 2R_n^T + n - 1 \le E \le 2R_n^* + n - 1. \qquad (5.9)$$

The ARG has $R = R_n^*$, whereas Hudson's algorithm has $R = R_n + R_n^T$. The expectation of R_n^T has not been found explicitly, but can be bounded upwards by

$$E(R_n^T) \le \rho(\rho + 1) \left(\sum_{i=1}^{n-1} \frac{1}{i} \right)^2. \qquad (5.10)$$

The bound is crude as can be seen by comparison with $E(R_n^*)$ which depends on $\sum_{i=1}^{n-1} 1/i$ only. For increasing n and fixed ρ, $E(R_n)$, $E(R_n + R_n^T)$, and $E(R_n^*)$ increase at similar rates, whereas for fixed n and increasing ρ, $E(R_n^*)$ increases at a much faster rate than $E(R_n + R_n^T)$. Thus, there is a huge gain in time spent on computation for large ρ in choosing Hudson's algorithm instead of the ARG. The spatial algorithm, in contrast, produces more events than given by the lower bound in equation (5.9). Simulation results show that the spatial algorithm becomes computationally heavy for large ρ, though it is not known whether E increases like e^ρ.

5.6 The effect of a single recombination event

Assume that only one recombination event has happened in the history of a sample and that the break point is in p. To the left of p there will be one local tree and to the right there will be another local tree. Two neighbour trees cannot be any two trees: one must be obtainable from the other by a transfer of a subtree within the rest of the tree (see Section 5.11.1). The relationship between neighbour trees falls into a few simple cases:

Type 1: The two recombining sequences coalesce before they coalesce with any other sequence in the sample (Figure 5.17). In this case the two genealogies will be identical and no method based on knowledge of the genealogies can detect that a recombination has occurred in the sequence history. Such events are effectively invisible. The ancestral material on a sequence was separated to be on two sequence, only to be joined again by a coalescent, before any of them had interacted with other ancestral sequences. A method that could detect this event would have to detect this physical linkage, for instance if the substitutional process involved coupling between positions far apart in the molecule.

Type 2: The unrooted tree topology remains the same but branch lengths change. This occurs when one of the two recombining sequences merges with one other sequence before coalescing with the other recombining sequence again (see Figure 5.18), or if there are only three ancestral lineages at the time of the event.

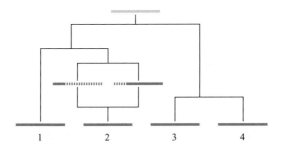

Figure 5.17 Genealogical history with one recombination resulting in identical tree topologies and branch lengths on each side of the recombination spot. Dark grey: ancestral material, light grey: The MRCA, dotted lines: non-ancestral.

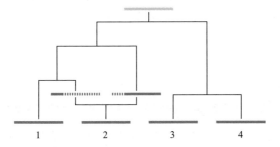

Figure 5.18 Genealogical history with one recombination resulting in the same tree topology, but different branch lengths on each side of the recombination spot. Dark grey: ancestral material, light grey: The MRCA, dotted lines: non-ancestral.

Type 3: The tree topologies change (see Figure 5.19). This can be defined negatively by being the remaining cases, but it might as well be defined by the number of sequences that merge with the recombining sequences before they coalesce with each other. In order for the topology to change two or more sequences must merge with the recombining sequences, before the two recombining sequences merge again.

Figure 5.20 illustrates the three types of events on a single tree. The probabilities of different categories of events can be calculated and are tabulated in Table 5.1. One surprising feature is the very high frequency of invisible recombinations for even a high number of sequences. This implies that methods trying to detect recombinations by detecting change in tree topologies will miss the majority of recombinations. The probability of an invisible recombination event will go to zero as the number of sequences increases, but very slowly.

These probabilities can be calculated using simple combinatorial arguments. The probability for an invisible recombination is the easiest. Assume the recombining sequences are created while there are $k \leq n$

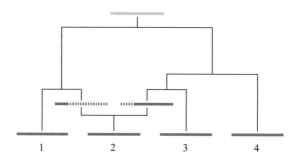

Figure 5.19   Genealogical history with one recombination resulting in different tree topologies on each side of the recombination spot. Dark grey: ancestral material, light grey: the MRCA, dotted lines: non-ancestral.

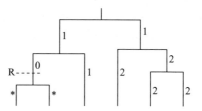

Figure 5.20   The genealogical consequences of recombination and recoalescence. If the recombination happens at the edge labelled 0, recoalescence to this edge will not change tree or branch lengths, recoalescence to the branches labelled 1 will give the same topology, but change some branch lengths. Recoalescence to the remaining branches will change the topology. It is impossible to recoalesce to the branches labelled '*', because they are more recent than the recombination event.

Table 5.1 Probability of different types of events[a]

n	P_n^1	P_n^2	P_n^3
2	0.333	0.667	0.000
3	0.297	0.703	0.000
4	0.272	0.655	0.073
5	0.256	0.616	0.134
6	0.243	0.574	0.183
10	0.212	0.488	0.300
15	0.191	0.435	0.374
500	0.098	0.232	0.670
∞	0	0	1

[a] P_n^i is the probability of a type i event in a sample of size n.

lineages. This has probability

$$q_k = \frac{1}{k-1} \left(\sum_{i=1}^{n-1} \frac{1}{i} \right)^{-1}. \tag{5.11}$$

Let p_{k1} be the probability that the two sequences carrying ancestral material to the recombinant sequence coalesce with each other before they coalesce with any other ancestral lineages. This satisfies the recursion (remember there are $k+1$ lineages just after the recombination event),

$$p_{k1} = \binom{k+1}{2}^{-1} + \binom{k-1}{2}\binom{k+1}{2}^{-1} p_{k-1,1}, \tag{5.12}$$

with $p_{21} = 1/3$. Either the two sequences coalesce at the first event—probability $\binom{k+1}{2}^{-1}$—or two of the remaining $k-1$ sequences coalesce—probability $\binom{k-1}{2}\binom{k+1}{2}^{-1}$. The last possibility that one of the two recombining sequences coalesce with one of the remaining $k-1$ sequences does not result in a type 1 event. Recursion (5.12) is solved by

$$p_{k1} = \frac{2}{3k}. \tag{5.13}$$

The probability of an invisible recombination (type 1 event) is then

$$P_n^1 = \sum_{k=2}^{n} p_{k1} q_k = \frac{2}{3} \left(1 - \frac{1}{n} \right) \left(\sum_{i=1}^{n-1} \frac{1}{i} \right)^{-1}. \tag{5.14}$$

A similar recursion can be written for the probability, p_{k2}, of a branch length change. The solution is

$$p_{k2} = \frac{8}{9k} + \frac{8}{(k+1)k^2(k-1)}\left(\frac{1}{3} + 4\sum_{j=1}^{k-2}\frac{1}{j}\right). \tag{5.15}$$

The probability, p_{k3}, of a topology shift follows from $p_{k3} = 1 - p_{k1} - p_{k2}$. Closed expressions for the probabilities, P_n^2 and P_n^3, of type 2 and 3 events, respectively, can be found similar to P_n^1. However, they do not reduce to simple expressions.

5.7 The number of recombination events

In this section we list some moments of variables that count different types of recombination events.

As stated in equation (5.6) the expected number of recombination events in ancestral material is

$$E(R_n) = \rho\sum_{i=1}^{n-1}\frac{1}{i}. \tag{5.16}$$

The expectation is linear in ρ such that histories of sequences of double length have double the number of recombination events on average. The variance of R_n can be expressed in terms of the correlation $f_n(x)$ between the total branch lengths of two local trees separated by $x/2$ recombination units (the whole sequence is $\rho/2$ units)

$$\mathrm{Var}(R_n) = \rho\sum_{i=1}^{n-1}\frac{1}{i} + 2\int_0^\rho(\rho - x)f_n(x)\,dx_n \tag{5.17}$$

(Hudson 1983a). The last term is for large ρ of order $\log(\rho)/\rho$, thus disappearing with increasing ρ. For $n = 2$, $f_n(x)$ is known:

$$f_2(x) = \frac{18 + x}{18 + 13x + x^2}.$$

The events that count in R_n can be divided into three subtypes (Section 5.6) according to whether

Type 1: The topologies and branch lengths are the same for local trees close to the break point (Figure 5.17).

Type 2: The unrooted topologies are the same, but branch lengths differ (Figure 5.18).

Type 3: The unrooted topologies differ (Figure 5.19).

In genealogies with many recombination events the type of a particular event is determined by the local trees in a small neighbourhood around the break points such that there are no other break points in that neighbourhood. Call the number of these events R_n^1, R_n^2, and R_n^3, respectively. The means of R_n^1, R_n^2, and R_n^3 are

$$E(R_n^1) = \frac{2}{3}\left(1 - \frac{1}{n}\right)\rho, \tag{5.18}$$

$$E(R_n^2) = \rho \sum_{i=1}^{n-1} \frac{1}{i} - E(R_n^1) - E(R_n^3), \tag{5.19}$$

and

$$E(R_n^3) = 16 \sum_{k=4}^{n} \frac{1}{(k+1)k^2(k-1)^2} \left\{ \sum_{i=2}^{k-2} \left[\frac{1}{i} \sum_{i=2}^{j} j^2(j+1) \right] \right\} \rho, \tag{5.20}$$

respectively (Hudson and Kaplan 1985, Wiuf et al. 2001). Eventually for large n most events are of type 3, but surprisingly many are of types 1 and 2 for small sample sizes. For large n,

$$E(R_n^3) \approx \left(1 - \frac{2.14}{\log(n)}\right). \tag{5.21}$$

5.8 The probability of a data set

There exists no set of recursions on a finite set of configurations like the recursions of Ethier and Griffiths (1987) that allows calculations of the probability of a data set in the presence of recombination. However, the recursions by Griffiths and Tavaré (1994) were generalised by Griffiths and Marjoram (1996) and can provide the basis of simulation algorithms:

$$
\begin{aligned}
&(n(n-1) + a\theta + b\rho)Q(\mathbf{A}, \mathbf{M}, \mathbf{n}) \\
&= n \sum_{1} (n_i - 1)Q(\mathbf{A}, \mathbf{M}, \mathbf{n} - \mathbf{e}_i) \\
&\quad + 2n \sum_{2} (n_k + 1 - \delta_{ik} - \delta_{jk})Q(\mathbf{A}, \mathbf{M}, \mathbf{n} - \mathbf{e}_i - \mathbf{e}_j + \mathbf{e}_k) \\
&\quad + \theta \sum_{3} (n_k + 1)Q(\mathbf{A}, \mathbf{M}(m), \mathbf{n} - \mathbf{e}_i + \mathbf{e}_k) \\
&\quad + \frac{\rho}{n+1} \sum_{4} \int (n_i + 1)(n_j + 1)Q(\mathbf{A}_k^{ij}(x), \mathbf{M}_k^{ij}(x), \mathbf{n}_k^{ij}(x))\, dx.
\end{aligned}
\tag{5.22}
$$

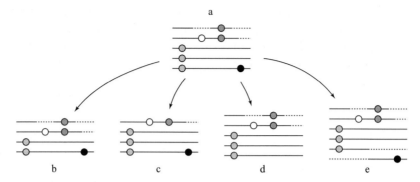

Figure 5.21   The present configuration (top of figure) can be created by four classes of events. Two identical sequences could coalesce, two sequences identical in common ancestral material could coalesce, a sequence can mutate and lastly a sequence could have arisen by recombination. This is here illustrated by taking one representative of the four classes, while in the recursion it is necessary to sum over all possibilities and in the case of recombination it is necessary to integrate over all recombination positions. a, b, c, d, and e correspond to the five lines in equation (5.22) in that order. In the example in the text, a is the sum of all dark lines, excluding dashed lines, and b is the sum of all dark lines, including the trapped dashed line in the top sequence.

This recursion is obtained by the usual backward–forward argument as in this case illustrated in Figure 5.21. **A** describes the distribution of ancestral material, **M** is the position of the mutations and **n** is the multiplicity of types, defined as in the case without recombination (see equation (2.27)). This notation can be illustrated on the 'a' configuration in Figure 5.21: **n** is $(1, 1, 2, 1)$, since the third and fourth sequences are identical and the remaining distinct. **A** is a list of intervals for the four distinct types, like $([0.0, 0.31], [0.47, 0.68]), ([0.0, 0.72]), ([0.0, 1.0]), ([0.0, 1.0])$, where the first type has two ancestral intervals, the second one interval and the last two types consist of ancestral material from start to end. **M** specifies mutations, for example, $0.18:(0, 0, 1, 1), 0.38:(-, 1, 0, 0), 0.58:(1, 1, 0, 0), 0.77:(-, -, 0, 1)$. The real number indicates the position, 0 the ancestral state, 1 the mutated state and '$-$' that the state is not specified, since it is in the non-ancestral material of that sequence.

δ_{ij} is one if i and j are identical, and otherwise zero; $\mathbf{M}(m)$ denotes removal of mutation m; $\mathbf{A}_k^{ij}(x)$, $\mathbf{M}_k^{ij}(x)$, and $\mathbf{n}_k^{ij}(x)$ denote the updating of **A**, **M**, and **n**, respectively, after a recombination event with break point x. A sequence of type k is broken up into two new types not previously in the sample, one of type i and one of type j, as shown in Figure 5.21. a is the sum of all ancestral material (it is only necessary to keep track of mutations in ancestral material), and b is the sum of all material, where recombination can happen (which includes trapped material). In the example $a = (0.31{-}0.0) + (0.68 - 0.47) + (0.72 - 0.0) + 1.0 + 1.0 = 3.24$ and $b = (0.68 - 0.0) + (0.72 - 0.0) + 1.0 + 1.0 = 2.40$.

$Q(\mathbf{A}, \mathbf{M}, \mathbf{n})$ is the probability or density of the $(\mathbf{A}, \mathbf{M}, \mathbf{n})$ configuration conditional on the positions of the mutations and the beginnings and ends of ancestral segments. The first sum is over all possible coalescence events of identical sequences, the second sum is over all coalescence of sequences with the same configuration of mutations in common ancestral material, the third sum is over all mutations of multiplicity one (singletons) and the last combined sum and integral is over all possible ways to generate a sequence by recombination. This recursion is initialised by $Q(\mathbf{A}, \mathbf{M}, \mathbf{n}) = 1$, where $\mathbf{n} = (1)$, $\mathbf{A}$ corresponds to one interval of ancestral material covering the complete sequence and $\mathbf{M}$ corresponds to no mutations. If the scaled recombination rate ρ is zero, the second and fourth terms can be ignored and the recursion reduces to the recursion in (2.27). This initialisation corresponds to waiting until the GMRCA, which is computationally slow. It can be initialised, when an ancestral sequence only has segments that are the ancestors to all the sequences in the sample.

This recursion is illustrated in Figure 5.21. The ancestral states of the mutations are assumed to be known. The present configuration 'a' has four mutations at four positions (infinite site assumption). Alleles with identical configurations of mutations and ancestral material can coalesce as shown in the 'b' configuration, where the third and fourth sequences from the present configuration have found a common ancestor. Sequences can also coalesce if they are identical on the ancestral segments as shown in the 'c' configuration, where the first and second sequences in the present configuration coalesce. The 'd' configuration illustrates the event of removing a mutation (the rightmost). The 'e' configuration shows a recombination event that splits the fifth allele in the present configuration into two new sequences. The recombination could have been anywhere in material spanned by ancestral material (also in trapped material) and all these possibilities must be integrated out.

The recursions for probability of data generated by histories with recombination are harder because they cannot be used to find the probability of the data even with infinitely much computing power as the number of ancestral states multiplies ad infinitum. Even if the number of ancestral configurations had been bounded, the necessity of integration in the last term makes the computations very hard.

5.9 The number of segregating sites

With and without recombination the expected numbers of S_n and η_i are the same, that is,

$$E(S_n) = \theta \sum_{i=1}^{n-1} \frac{1}{i}, \tag{5.23}$$

and

$$E(\eta_i) = \frac{\theta}{i}, \qquad (5.24)$$

for $i = 1, \ldots, n - 1$. However, the variances changes, for example, the variance of S_n becomes

$$\mathrm{Var}(S_n) = \theta \sum_{i=1}^{n-1} \frac{1}{i} + \frac{2\theta^2}{\rho^2} \int_0^\rho (\rho - x) f_n(x) \, dx, \qquad (5.25)$$

where $f_n(x)$ is defined as in equation (5.17).

The variance attains the largest value for $\rho = 0$ and decreases towards $\theta \sum_{i=1}^{n-1} 1/i$ with increasing ρ. The reason behind this is that with increasing ρ, S_n sums mutations over many (almost) independent trees, thus reducing the variance.

5.10   The coalescent with gene conversion

Wiuf and Hein (2000) and Wiuf (2000a) included simple models of gene conversion into Hudson's model of recombination. It is assumed that the time to a gene conversion event is exponentially distributed (in the continuous time approximation) with a parameter $\gamma = 4Ng$, where g is the probability that a gene conversion tract initiates within a sequence in one generation.

The tract length in nucleotides is drawn from a specified distribution: Empirical evidence points to a geometric distribution with parameter $q > 0$. In the infinite sites approximation the tract length becomes exponentially distributed with intensity $Q = qL$ such that $1/Q$ is the mean length of the gene conversion tract. If a gene conversion event occurs, the first break point is chosen uniformly on the sequence, and the second break point is chosen a distance away from the first as determined by a random number from the tract length distribution. The upper part of Figure 5.22 shows how a gene conversion event distributes the ancestral material on two different ancestors. Note that a gene conversion event may only have one break point within the sequence if the tract extends beyond the end of the sequence, or if a tract initiates outside the sequence but ends within. Therefore some events will be indistinguishable from recombination events.

The time to the next event for a set of k sequences is exponentially distributed with parameter

$$\frac{k(k-1)}{2} + \frac{\rho}{2}k + \frac{\gamma Q^*}{2}k, \qquad (5.26)$$

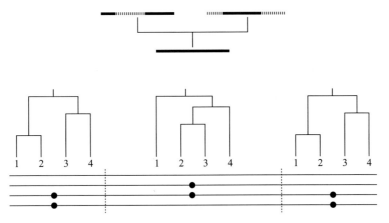

Figure 5.22 A history with gene conversion. The top figure shows the effect of a gene conversion in a sequence: The ancestral material is spread on two sequences such that the flanking regions are on one sequence, the middle part on another sequence. The bottom figure shows how the tree changes over the sampled sequences. The two outer trees are the same. This interpretation in terms of genealogies also leads to a test of gene conversion in terms of compatibilities. It is illustrated lowest in the figure: three informative sites are found that are placed such that sites 1 and 3 are compatible, while the middle site is incompatible with both 1 and 3. This pattern is considerably more unlikely under a pure recombination model than if gene conversion is allowed.

where

$$Q^* = 1 + \frac{1}{Q}(1 - e^{-Q}) \tag{5.27}$$

is a factor that corrects for the possibility that a tract initiates outside the sequence. If Q is small, the tract is long and Q^* is approximately equal to two. In that case pure gene conversion looks like recombination at rate $\rho = 2\gamma$. If Q is large, the tract is small and Q^* is almost one. In that case a tract is just a tiny spot on a sequence. As a consequence the genealogy looks like a coalescent genealogy where all sites share history apart from a few sites (the spots) that have different histories.

The probability that the next event is a gene conversion event is

$$\frac{\gamma Q^*}{k - 1 + \rho + \gamma Q^*} \tag{5.28}$$

with similar probabilities for the other events. Figure 5.23 shows how the location of the starting point of the tract depends on whether one or both end points are within the sequence.

Figure 5.22 also shows the consequence of a gene conversion event on the genealogy relating the sampled sequences. Scanning the sequences from left to right, the start of the gene conversion will look like a

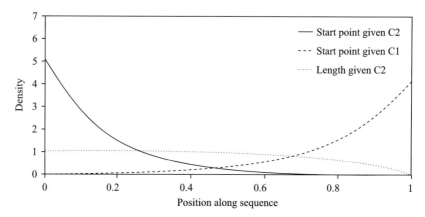

Figure 5.23 The probabilities of the gene conversion tract covering an interval. C2 is the event that the gene conversion tract is contained in the interval and C1 the event that it is not. The almost flat curve is the density of the length of the tract given C2. The strongly descending curve is the density of the start position given C2. The strongly ascending curve is the density of the start position given C1.

recombination—taking a subtree and moving it. The end of the gene conversion will look like the reverse of the start of the gene conversion—taking the subtree and moving it back again. Obviously this creates a change over short distances, but not over long distances.

5.11   Gene trees with recombination—from incompatibilities to minimal ARGs

Genealogical histories without recombinations are much easier to represent than genealogies with recombinations. This section discussed some purely combinatorial and representational questions that are encountered, when analysing data subject to recombination. A proper statistical treatment of these issues belongs to the future.

First, the phylogenetic consequences of recombination are discussed, followed by some considerations and examples of the difficulty in reconstructing recombinations. Then three central statistics of the data are discussed: The minimal number of trees compatible with the sample, which also is a lower bound to the number of recombination events in the sample's history, an improved lower bound to the number of recombination events, and lastly the minimal ARG describing the data.

Reconstructing—in a classical phylogenetic sense—a full history of a set of sequences is close to impossible. The complexities introduced by recombination enters at several levels. Within the infinite site model recombination can be deduced to has occurred between two positions if they are incompatible. However, this only states that at least one recombination has occurred between them. If the actually genealogies at the two sites were known more recombinations could have occurred. Lastly, as we have seen, recombinations often do not change the genealogy, so even full knowledge of all local trees for a set of sequences would underestimate the true number of recombinations events.

5.11.1 Recombination as subtree transfer

Hudson's model of recombination emphasises how genetic material is transmitted from parent to offspring when visualised in time reversed perspective. The spatial algorithm illustrates another important way of thinking about recombination: Recombination as local modification of genealogical history. In Figure 5.14 a recombination event is made by cutting a branch and moving the subtree hanging under the branch to a different location. This operation is called *subtree transfer* or *subtree regrafting*. Thus the basic idea in the spatial algorithm is to build the genealogy or graph by subtree transfers while taking probabilistic constraints imposed by the coalescent process into account. The time when the recombination happens (that is where the subtree is cut) must be more recent than the time when the subtree is attached to the graph again. Also the local tree at the recombination break point can be linked to earlier local trees through trapped material. When comparing trees, the latter constraint is of less importance.

One way of comparing two local trees would be by counting the minimum number of subtree transfers required to change one tree into the other. Now, the minimum number of transfers depends on how much information is included in the description of the tree. At the lowest level one could record just the unrooted tree topology, or the topology including the root which implies a relative age-ordering of coalescent events as one moves from the root towards the leaves. Finally, one could record a relative age-ordering of all events irrespective of whether they occur on the same path from the root to the leaves or not (a total ordering), or a complete specification of the actual times of all events. Note that a tree with specification of branch lengths is equivalent to the latter: The distance from the root to the leaves is the same for all leaves; thus halfway between the leaves separated by the longest distance is the location of the root. In a similar way all other coalescent events become dated absolutely. A coalescent topology corresponds to the tree with a total age-ordering of events and a tree with all events dated corresponds to a realisation of the coalescent process.

Figure 5.24 illustrates these distance measures of two simple genealogies with six leaves. Obviously, a path of subtree transfers converting the first genealogy into the second genealogy can be converted into a path of the same length converting the second into the first.

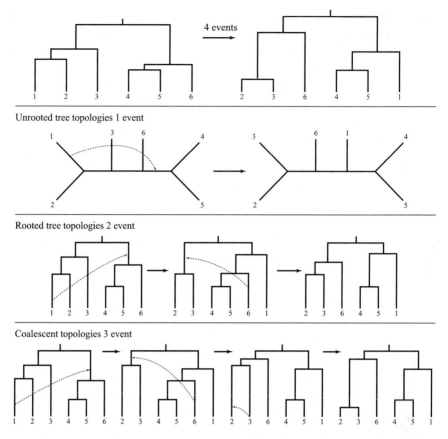

Figure 5.24 Various distance measures on trees and topologies. The unrooted topologies are one subtree transfer away, the rooted two, the age-ordered three, and the trees with full specification of times four transfers away.

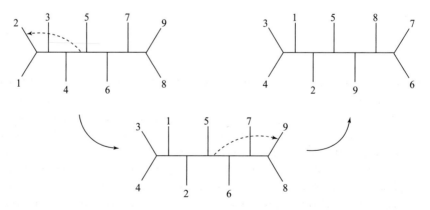

Figure 5.25 An example of where the two distance measures differ. The two topologies are two subtree transfers away. The second violates the time constraint imposed by the coalescent process. Example by Thomas Christensen.

If the genealogies are converted into two unrooted topologies, then the first can be converted into the second by one subtree transfer—by moving the little tree only containing leaf 1 over to sit outside $(4,5)$ (see second highest panel of the figure). It is possible to assign ages to all internal nodes and to the roots, and then the resulting genealogies would be a subtree transfer from each other. But these two genealogies would not be the same as the two original genealogies.

If the information concerning the root is retained, that is, giving rooted tree topologies, then these will be two subtree transfers apart (see second-lowest panel of the figure). The single operation used on the unrooted topologies would have six sitting on the wrong side of the root. This can be rectified by an extra transfer.

If the internal ordering and the root is kept, that is, coalescent topologies are considered, then three subtree transfers are needed (see lowest panel of the figure). The internal nodes might now be labelled $(7–10)$, so the lower numbered labels are the more recent. If the two transfers are used from the rooted topologies above, then internal nodes 9 and 10 have not been changed. The resulting coalescent topology has the right unrooted topology, but the age ordering of 9 and 10 are wrong relative to the target coalescent topology and a subtree transfer has to be postulated to move the ancestor of 2 and 3 further down.

If the information in the complete genealogies is kept, and for instance the dates of the roots are different, then a subtree transfer would be necessary to move the root to the correct time (top panel).

In these different cases there is in general no reason to believe that the best path (a series of transfers) of the simpler problem is always a part of the path of the more complicated problem.

In the coalescent with recombination the number, ζ_i, of events that involve a transfer of a subtree of size i has mean

$$E(\zeta_i) = \rho \frac{1}{i}, \tag{5.29}$$

for $i = 1, \ldots, n - 1$. Thus it is more likely that small trees are moved rather than large trees.

When dealing with sequence data the situation is different, since local trees cannot be observed. If a number of SNP polymorphisms is observed each SNP polymorphism corresponds to a partition of the sequences into two subsets. All that can be said about the genealogy underlying the polymorphism is that the topology belongs to a certain class of topologies, namely the class of topologies that all have the given partition. This implies that even less can be said about the coalescent topology that also includes an age-ordering of all coalescent events in the history of the SNP. If two topologies, T_1 and T_2, or two classes of topologies, each corresponding to a SNP, are compared, one has to take into account that each subtree transfer applied to change T_1 into T_2 imposes constraints on the age-ordering. Otherwise there would not be a coalescent topology (or a realisation of the coalescent process) that is in agreement with the series of subtree transfers. Figure 5.25 provides an example of two topologies where the number of subtree transfers differ according to whether age-ordering constraints are taken into account or not. A subtree can only be moved in its root and thus two subtree transfers could impose conflicting constraints. In Figure 5.25 two trees with nine leaves represented as unrooted tree topologies are two subtree transfers away from each other. But if one tries to pick coalescent topologies that would correspond to these events, it is impossible because the two subtree transfers give conflicting information about the subtrees that must exist in the coalescent topology. In most cases, and for small number of sequences, the two ways of counting the distance between topologies agree.

If an infinite sites model is assumed, the sequences cannot be represented by a gene tree if all four combinations of 0 and 1 are present in two columns. At least one recombination event is required to explain these two columns. But what is the least number, R_M (M is for minimum), of events required to explain the whole sample? What is the least number, T_M, of gene trees required to explain the sample? Are the two numbers related? In general, $R_M \geq T_M - 1$, because there must at least be one recombination event between any two trees (if there were zero events between two trees they could be collapsed into a single tree). Further, each site is compatible with a gene tree so at most S_n gene trees are required to explain the whole sample, that is, $T_M \leq S_n$. The number of recombination events is also bounded upwards: $R_M \leq (n-1)(S_n - 1)$. If $n-1$ of the sequences are each split into S_n fragments (each fragment is a single nucleotide) using $(n-1)(S_n - 1)$ recombination events, a history that is compatible with the infinite sites assumption can easily be constructed. A simple example is shown in Figure 5.26. This bound is in general a crude overestimate.

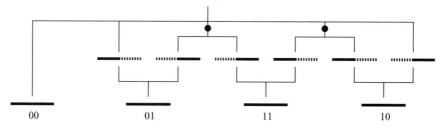

$$00 \qquad 01 \qquad 11 \qquad 10$$

Figure 5.26   There is always a history with at most $(n-1)(S_n - 1)$ recombination events that explains the sample. In the example $n - 1 = 3$ of the sequences are each broken into $S_n = 2$ pieces that subsequently can coalesce to the unbroken sequence.

The actual physical order of the segregating sites is important. This order was irrelevant when constructing a gene tree without recombination, because all sites share the same history. Consider the following two samples of four sequences and three segregating sites,

1. 0 0 0   0 0 0
2. 0 1 0   0 0 1
3. 1 0 1   1 1 0
4. 1 1 1   1 1 1.

Obviously, the first sample requires two recombination events, whereas the second only requires one; this despite the fact that one of the samples (either one) can be constructed from the other by permuting two columns:

       A B C D   A B C D
1. 1 0 1 0   1 0 1 1
2. 1 0 0 1   1 0 0 0
3. 0 0 0 0   0 0 0 0
4. 0 1 0 1   0 1 0 0
5. 0 1 1 0   0 1 1 1.

The above examples have the same pairs of incompatible sites: (A, C), (A, D), (B, C) and (B, D), whereas (A, B) and (C, D) are compatible. In both examples we infer that $T_M = 2$ because sites A and B can be explained by one tree, T_1, and sites C and D by another tree, T_2, incompatible with the first tree. However by trial and error we conclude that two recombination events are needed to transform T_1 into T_2, that is, $R_M = 2$, whereas only one event is needed in the second example, that is, $R_M = 1$. Since a single recombination event is a transfer of a subtree Figures 5.27 and 5.28 are helpful. It is easily seen that the left tree in Figure 5.28 cannot be created from Figure 5.27 with a single transfer.

This example shows that R_M cannot be found from T_M and vice versa, but also that R_M cannot be found from the list of incompatible pairs of sites. As a further complication, the two recombination events in the first example can either happen between sites B and C, or one of them between sites B and C, the other between A and B. In other words, in general there

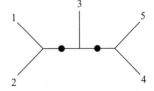

Figure 5.27   The figure shows the tree induced by the first two columns. The patterns corresponding to the third and fourth column cannot be placed on this tree with less than two mutation events per column.

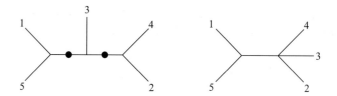

Figure 5.28   The left tree corresponds to columns 3 and 4. This tree cannot be obtained from the tree in Figure 5.27 with a single subtree transfer. To the right is the tree for the last two columns in the second example. It can be obtained by moving sequence 1 next to sequence 5 in Figure 5.27. The two examples have the same configurations of compatible sites, but need different numbers of recombination events.

is not a unique way to represent histories with R_M events graphically. The same is true for T_M, as the following example shows,

```
    A B C D E
1.  0 0 0 0 0
2.  0 0 1 0 1
3.  1 0 0 0 1
4.  1 0 1 0 0
5.  0 0 0 0 1
6.  0 1 0 0 0
7.  0 1 0 1 0.
```

Here, $T_M = 3$ and three trees can be chosen in three different ways: (a) $T_1 = \{A, B\}, T_2 = \{C\}$, and $T_3 = \{D, E\}$, (b) $T_1 = \{A, B\}, T_2 = \{C, D\}$, and $T_3 = \{E\}$, or (c) $T_1 = \{A\}, T_2 = \{B, C\}$, and $T_3 = \{D, E\}$.

In contrast to R_M, T_M is easy to find, for example, Hudson and Kaplan (1985) provide an algorithm based on the list of pairwise incompatible sites. Conversely, R_M does not seem to be easy to find. One can apply a parsimony algorithm to find R_M but some sort of exhaustive search seems to be necessary (Hein 1990; Wiuf 2002). One drawback of such an algorithm is that it is extremely time consuming which makes it impractical for large data sets, $n > 10$.

5.11.1.1   How to find T_M

Hudson and Kaplan (1985) provide one algorithm to find T_M. Wiuf (2002) provides a different algorithm that gives the same result. Here Wiuf's algorithm is shown because it is easier to see that the algorithm produces the desired result. There are several steps.

1. Calculate the matrix, $D = \{D_{ij}\}, i, j = 1, \ldots, S_n$, of pairwise compatibilities. That is, for each pair of sites, i and j, determine whether all four gametes, 00, 01, 10, and 11, are present in the sample. If yes, put $D_{ij} = 1$ and otherwise put $D_{ij} = 0$.
2. Set a counter, c, equal to 1.

3. Find the smallest $c < j$ such that $D_{ij} = 1$ for some $c \leq i < j$. The site j is the first site that is incompatible with another site, $i \geq c$.

4. Record the interval $[c, j - 1]$. Set $c = j$ and repeat 3.

5. The recorded list of intervals is a list that can be explained by T_M trees.

This algorithm scans the sequences left to right and only switches to an island with a new tree, when the appropriate incompatibility is encountered.

5.11.2   Recombination inferred from haplotypes

R_M and T_M are two summary statistics. Because R_M is computationally difficult to find, T_M is frequently used. However it is less informative because $R_M \geq T_M$. In this section a third summary statistic is introduced: the haplotype statistic H_M (Myers and Griffiths 2003). Recombination increases the expected number of haplotypes because two different haplotypes can create a third (new) haplotype without assuming further mutations. For example without recombination and only two segregating sites there can be at most three different haplotypes, whereas there can be four with recombination. Mutations cannot explain all four haplotypes. Consider a tree relating six sequences, Figure 5.29.

If there are no mutations in the tree all six sequences are identical and there is just one haplotype. Then $1 = K_n \leq S_n + 1 = 1$, where K_n is the number of different haplotypes. As more mutations are put onto the tree more haplotypes are created. Each new mutation in the tree can increase the number of haplotypes by at most one. This we illustrate by example: Assume three mutations have been placed (left part of the figure). Then there are four distinct haplotypes and $4 = K_n \leq S_n + 1 = 4$. If the next mutation is at A one haplotype group (sequence 2, 3, and 4) is split into two groups (sequence 2, and 3 and sequence 4); then there are four different haplotypes and $5 = K_n \leq S_n + 1 = 5$. If the mutation is at B no new haplotypes are created because there is already a mutation in B's branch. Finally, if the new mutation is at C the number of haplotypes also remains the same. The

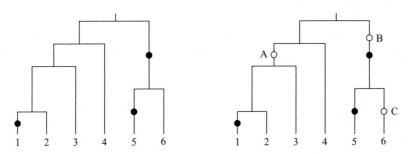

Figure 5.29   In the left tree each mutation enlarges the number of haplotypes, while on the right tree, only the extra white mutation at A does not create a new haplotype.

two mutations in the lineage of sequence 5 already define sequence 5 and 6's haplotypes. In consequence, each mutation splits at most one haplotype group into two, thereby increasing the number of haplotypes by at most one.

This observation can be formulated in the following way: If there are S_n segregating sites and no recombination then there are at most $S_n + 1$ different haplotypes. If there are K_n different haplotypes, then at most $S_n + 1$ haplotypes can be explained by mutations without invoking recombination. Consequently, the rest of the haplotypes, $K_n - S_n - 1$, must be due to recombination. Consider another example:

	A	B	C	D	E	F
1.	0	0	0	0	0	0
2.	0	0	0	1	0	0
3.	0	0	1	0	0	0
4.	0	0	1	1	0	0
5.	1	0	0	0	0	0
6.	1	0	0	1	1	0
7.	1	1	1	0	0	1
8.	1	1	1	1	1	1.

There are eight sequences, eight different haplotypes, and six segregating sites, A−F. Using the formula gives $K_n - S_n - 1 = 8 - 6 - 1 = 1$ and there must be at least one recombination event in the sample's history. However, the same question could be asked for any subset of the sites. For example, A, C, and D define all eight haplotypes and thus require $8 - 3 - 1 = 4$ recombination events. Now, a history of A−F is also a history of a subset of the sites, in particular of the sites A, C, and D, and thus there must be at least four recombination events in the history of the whole sample. The same could be done for the subset (E, F). Here $4 - 2 - 1 = 1$, so one recombination event is required. Since the two subsets (A, C, D) and (E, F) define non-overlapping regions of the sequences, it must be that a history of A−F requires at least $4 + 1 = 5$ recombination events. The same could not have been concluded if (B, E) was chosen instead of (E, F). At least one event is required in the history of (B, E), but the region defined by (B, E) is overlapping with that defined by (A, C, D) and therefore, a recombination event in the history of (B, E) might also be part of the history of (A, C, D).

5.11.3 From local to global bounds

The informal definition goes like this: Partition the sequences into non-overlapping regions. Calculate the bound $K_n - S_n - 1$ for each region, perhaps excluding some sites from the region. Choose the partition that gives the highest number. That number is H_M. In the example there were two regions A–D and D–F. In the first site B was left out, in the second D. Note that $T_M = 4$, because A and C are incompatible, so are C and D, and E and F. It is always true that $H_M \geq T_M$.

5.11.3.1   *How to find H_M*

The algorithm provided here is from Myers and Griffiths (2003). Define H_{ij} by

$$H_{ij} = \max\{b_{ik} + H_{kj} \mid k = i+1, \ldots, j-1\} \qquad (5.30)$$

and

$$H_M = H_{1S_n},$$

with boundary conditions $H_{ii} = 0$, $H_{i,i+1} = b_{i,i+1}$, and $b_{ij} = 1$ if sites i and j are incompatible, and 0 otherwise. H_{ij} is the bound obtained for the regions spanned by the site i and j. For large data sets the algorithm can be quite slow because many partitions have to be tried out to find the optimal combinations of sites. This algorithm creates larger and larger intervals with a maximal combination of bounds on small segments and smaller intervals with maximal bounds on recombination events.

5.11.4   Minimal ARGs

One of the most famous data sets in the history of molecular popula-tion genetics was published by Martin Kreitman in 1983. It contained 11 sequences each about 3200 bp long of 11 alleles of ADH genes from *Dro-sophila melanogaster*. It had forty-three segregating sites. Twenty-eight of these are informative and could contain information in parsimony sense about recombination events. Determining the number of recombination events in this data set has been the motivation for at least four methods: A paper by Hudson and Kaplan from 1985 that produced 5 as a lower bound on the necessary recombinations events in the history of these data. Using Myers and Griffiths' H_M gave a lower bound of 6. Song and Hein (2003) produced a lower bound of 7 and subsequently Song and Hein (2004a) proved that this bound can be realised in an ARG and that no higher lower bound is possible. This minimal ARG is shown in Figure 5.30.

   The method obtaining this minimal ARG is very slow indeed and cannot be used on data sets of more than 9–10 sequences and is thus not practical for large data sets at present.

   The method scans the sequences and at each column keeps track of the cost of the minimal histories ending in all possible coalescent topologies in that column. Let $d(T_1, T_2)$ be the smallest number of recombination events needed to convert T_1 into T_2, and $s(T, i)$ be the number of substitutions

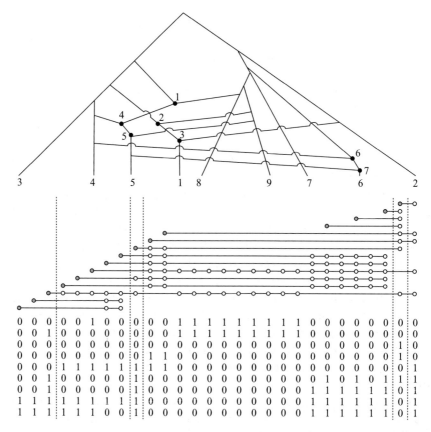

```
0 0 0 | 0 0 1 0 0 | 0 0 1 1 1 1 1 1 1 1 1 1 0 0 0 0 0 0 | 0 | 0
0 0 1 | 0 0 0 0 0 | 0 0 1 1 1 1 1 1 1 1 1 1 0 0 0 0 0 0 | 0 | 0
0 0 0 | 0 0 0 0 0 | 0 0 0 0 0 0 0 0 0 0 0 0 0 0 0 0 0 0 | 1 | 0
0 0 0 | 0 0 0 0 0 | 0 1 1 0 0 0 0 0 0 0 0 0 0 0 0 0 0 0 | 1 | 0
0 0 0 | 1 1 1 1 1 | 1 1 1 0 0 0 0 0 0 0 0 0 0 0 0 0 0 0 | 1 | 1
0 0 1 | 0 0 0 0 0 | 1 0 0 0 0 0 0 0 0 0 0 0 1 0 1 0 1 0 | 1 | 1
0 0 1 | 0 0 0 0 0 | 1 0 0 0 0 0 0 0 0 0 0 0 1 1 1 1 1 1 | 0 | 1
1 1 1 | 1 1 1 1 1 | 1 0 0 0 0 0 0 0 0 0 0 0 1 1 1 1 1 1 | 0 | 1
1 1 1 | 1 1 1 0 0 | 1 0 0 0 0 0 0 0 0 0 0 0 1 1 1 1 1 1 | 0 | 1
```

Figure 5.30 The lowest part is the Kreitman data with only informative sites shown, then identical sequences represented by one copy and segregating sites recoded as 0 and 1. The middle part indicates which nucleotides are incompatible—a pair with a dark ball to the left and an open ball to the right in the same line are incompatible. A recombination event must occur between two incompatible pairs, but obviously a recombination can explain several incompatible pairs. The top part is the minimal ARG explaining the sequences. From this ARG can be extracted genealogies that describe different segments of the full sequences as defined by the vertical lines. There are seven recombination events that could define eight segments, but they only define six segments as some recombination events occur between the same neighbouring pair of incompatible sites. The genealogy of site i can be recovered by starting in the bottom of this ARG and going left at recombination node k, if $i \leq k$, otherwise choose the left edge upwards from the recombination node. (Adapted from Song and Hein 2003)

needed to explain column i using tree T. When using the infinite site model, $s(T, i)$ will be infinity if more than one substitution is needed and $s(T, i) = 0$ otherwise. With this information a dynamical programming algorithm can be formulated as

$$W(T_2, i) = \min_{T_1}\{W(T_1, i - 1) + d(T_1, T_2) + s(T_2, i)\}, \tag{5.31}$$

and

$$R_M = \min_T \{W(T, S_n)\}. \tag{5.32}$$

$W(T, i)$ is the minimum number of recombination events required to explain the first i sites if the tree in site i is T.

The minimisation has to be over all T_1s at the $(i-1)$th column and the recursion above will give a value for all possible T_2 at column i. The recursion is initialised by $W(T, 1) = s(T, 1)$. The initialisation states that the cheapest history for the first column given a given coalescent topology, is just the cost of that column using that coalescent topology. The recursion states that the cheapest history for the first i columns given that the relationship of the sequences in column i is T_2 must be the optimal combination of the history up to the $(i-1)$th column and the cost of adding the ith column. This cost has both a substitution cost from explaining column i using T_2 and the recombination cost of transforming T_1 into T_2.

This minimisation algorithm has some resemblance to the spatial coalescent–recombination algorithm, but there is a crucial difference. The cost of changing to a new coalescent topology only depends on the coalescent topology in the previous column, while the corresponding probability would depend on the complete ARG for all the columns before the present column.

Applying this recursion will give a minimal set of coalescent topologies that can be combined into a single ARG. Figure 5.30 shows the above algorithm applied to the Kreitman data set, when no double mutations are allowed in any positions, which accelerates the algorithm. Recursion (5.31) has $T_M - 1$ as outcome if $d(T_1, T_2)$ is defined as $d(T_1, T_2) = 1$ if T_1 and T_2 have different topologies and 0 otherwise.

5.11.5   Topologies, recombination, and compatibility

Genealogies with recombinations pose challenging problems. The combinatorics of trees has been much studied, while similar efforts have not yet been undertaken for cases involving recombination.

It is obvious from the discussion of minimal ARGs, compatibility and the probability of tree topology changing recombination events, that detecting individual recombination events by inspection of the sample will only reveal a small fraction. Figure 5.31 show two sets of simulations for eight sequences with $\theta = 15$ and $\theta = 40$. This illustration shows the inherent difficulty in reconstructing recombination events. The dashed and solid lines are the expected number of recombination events and topology changing recombination events as a function of ρ. The lower set of *, ○, and ■ are the events recovered by the minimal ARG method, H_M and T_M, respectively for simulations using $\theta = 15$. The upper set are from simulations using

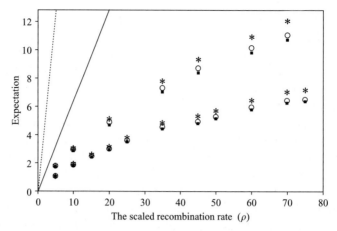

Figure 5.31   The dashed and full lines show the expected number, $E(R_n)$, of recombination events and the expected number, $E(R_n^3)$, of topology changing recombination events, respectively, for $n = 8$. The *, ○, and ■ symbols show the performance of the minimal ARG method (R_M), the haplotype bound (H_M) and the Hudson–Kaplan bound (T_M) respectively, for $\theta = 15$ (lower curves) and $\theta = 40$ (upper curves). The discrepancy between reconstructed and expected number of recombination events is very large indeed. The larger the mutation rate, the larger the fraction of reconstructed recombination events. As θ goes to infinity one would expect that all topology changing recombination events would be detected. (Adapted from Song and Hein 2004a.)

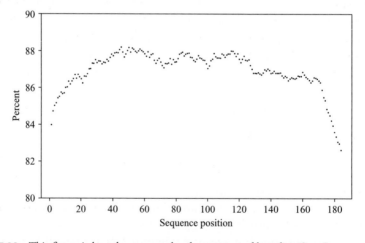

Figure 5.32   This figure is based on a sample of sequences of length $\rho/2 = 5$, (corresponding to 180 nucleotides) and 2000 simulations. In addition to the arch-shaped form due to better phylogenetic information in the middle of a sequence relative to the flanking regions, there is an asymmetry that could be caused by the dynamical algorithms choice to place necessary recombination events as rightward as possible. (Adapted from Song and Hein 2004a.)

$\theta = 40$. It is immediately seen that the recovered recombination events is only a small fraction of the total number of recombination events and also of the number of visible recombination events. It is also intuitive that a higher number of mutations will allow the detection of more recombination events.

The conclusions from investigations trying to reconstruct recombination events are thus negative, which is of course unfortunate. However, if the goal is to find the genealogical relationship between the sequences, the possibilities are better. First, a reasonable measure of similarity of genealogical histories needs to be defined. Comparing ARGs is an undeveloped topic relative to comparing classical phylogenies. If the local trees of a reconstructed ARG are compared with the local trees of the true ARG, comparing ARGs reduces to comparing trees. In Figure 5.32 one such comparison is shown for seven sequences with $\rho = 10$ and $\theta = 75$. The similarity of two trees is measured as the percentage of identical bipartitions they induce. Unrooted trees with seven leaves will have four such bipartitions corresponding to four internal edges. Again, if θ grew arbitrary large, this score would converge to 1.0. But for this simulation the average score along the sequence is around 0.85, which means that the local tree on average would have all or three bipartitions correct out of four possible. The shape of this score along the sequence is interesting as well and is readily interpretable: The local trees in the middle have more mutations to guide it towards the right tree, while at the borders of the sequence, there will only be mutations to one side.

In the approaches discussed, compatibility is a very simple and useful concept, that unfortunately catches too little of the underlying tree structure, while the method used to find the minimal ARG above seems very brute force and exhausting in testing all trees at informative positions that are compatible with that position.

Recommended reading

Griffiths, R. C. and Marjoram, P. (1997), An ancestral recombination graph, in P. Donnelly and S. Tavaré, eds., 'Progress in Population Genetics and Human Evolution', Springer Verlag, pp. 257–270.

Hudson, R. R. (1983a), 'Properties of a neutral allele model with intragenic recombination', Theor. Popul. Biol. 23(2), 183–201.

Hudson, R. R. and Kaplan, N. L. (1985), 'Statistical properties of the number of recombination events in the history of a sample of DNA sequences', Genetics 111(1), 147–164.

Lewin, B. (2003), Genes VIII, Oxford University Press.

6 Getting parameters from data

6.1 Introduction

One strength of the coalescent approach is that it is data-oriented with emphasis on the sample of sequences that the investigator has collected. It aims at describing the processes that generated the data, retrospectively describing the sample in terms of possible previous events. In contrast, pre-coalescent population genetics is typically prospective: given the present what is the likely outcome of evolution? In the previous chapters we have shown how it is possible to simulate and predict quantities under the coalescent. In particular we have seen that it is possible to simulate data sets under various conditions and with different parameters, and that it is possible to calculate the probability of a data set for a fixed set of parameter values. This situation is depicted in Figure 6.1. The evolutionary parameters (collectively referred to as ψ) can generate a variety of data where a given data set is just a single point in the possible space of data sets for a given ψ.

The typical investigator collects data with the aim of learning about the evolutionary processes and parameters that have shaped the data. Different data sets tend to have different probabilities of being generated for different values of ψ, and a given data set therefore provides information about ψ. Extracting information about the underlying processes and parameters are called statistical inference. One important aspect of statistical inference is *estimation theory*. It is the topic of the current chapter.

Figure 6.1 is in a sense the reverse of Figure 6.2. For a given data set we want to find a value $\hat{\psi}$ of ψ that best (in some sense) explains the data. This value is said to be an estimate of ψ, and a function assigning an estimate $\hat{\psi}$ to any possible data set is called an estimator of ψ. Data sets that are generated with the same parameter value ψ might provide very different estimates of ψ.

Figure 6.1 The evolutionary parameters determine which data may be generated with what probability.

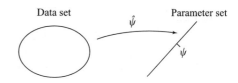

Figure 6.2   Inference on the parameter ψ from the data through the estimator $\hat{\psi}$.

Historically, statistical inference under coalescent has been based on summaries of data, for example, Watterson's estimator of the scaled mutation rate or the sample heterozygosity. Summary statistics are in general easily computed and the outcome of a test based on summary statistics is in general just as easily evaluated either from comparison with tabulated values or from simulation. However, performing inference from summary statistics risks discarding much of the information in the full data that otherwise could prove useful for inference about the underlying data generating process. For example, Watterson's estimator relies solely on the number of segregating sites in the sample and not on the actual pattern of mutations observed in the data.

In contrast, approaches that make full use of the data are often computationaly slow and difficult to implement in computer programs. Sometimes a method might even give unreliable results because it attempts to compute a quantity, say, the likelihood of the data, that is difficult to compute using the chosen approach. For example, using recursion (2.27) in Chapter 2 it is straightforward to calculate the probability of a given data set. However for realistic sample sizes the number of terms in the recursion blows up quickly and any program implementing recursion (2.27) cannot produce a result in a reasonable amount of time. If the task is to find the maximum likelihood estimate of θ and not just the likelihood for a single value of θ the problem is even harder: In principle, it involves calculating the probability of the data for many different parameter values and selecting the one that yields the highest likelihood of the data.

In this chapter we focus on estimation theory in relation to the coalescent. The stochastic nature of the coalescent process complicates parameter estimation because genes in a sample are dependent. We discuss some of these complications in the light of moment and pseudo-likelihood estimators of θ and ρ. In addition we discuss some advanced techniques for calculating the likelihood of a data set: Importance Sampling (IS) techniques and Markov Chain Monte Carlo (MCMC) methods. These techniques are essential for evaluating the likelihood of the data and thus for computing the maximum likelihood estimate.

6.2   Estimators of θ

This section concentrates on inference on the scaled mutation rate, θ. In neutral models θ plays a role different from the other genetic and demographic parameters that have been introduced so far. Parameters like ρ or β shape the genealogy of the sample; in contrast, θ does not shape the genealogy, it modifies genetic types. As a consequence, mutation-rich samples allow for more accurate estimation of 'genealogical shape'-parameters than do mutation-poor samples, simply because the information available about the sample's genealogy increases with the number of segregating sites.

One extreme situation is if mutation is impossible ($\theta = 0$). Consequently all samples have zero segregating sites and any reasonable estimator of θ is constantly 0 (e.g. Watterson's estimator). Despite the fact that θ is estimated with high accuracy (we always report the true value) there is no information in a sample about ρ or β. At the other extreme, θ large ($\theta \approx \infty$), the length of a branch is proportional to the number of mutations and we have full information about the genealogy up to a scaling factor.

6.2.1 Watterson's estimator

Watterson's estimator is one of two very popular estimators of the scaled mutation rate. The other is *Tajima's estimator*. They are both intuitively appealing in that they relate to easily computable summary statistics and have relatively simple interpretations.

Watterson's estimator, $\hat{\theta}_W$, is defined as

$$\hat{\theta}_W = \frac{S_n}{a_n}, \tag{6.1}$$

where $a_n = \sum_{j=1}^{n-1} 1/j$ (see also Section 2.5.4). Obviously, only the actual observed number of mutations, S_n, is used to calculate $\hat{\theta}_W$, not the sample configuration, and S_n is just compared to the average total branch length of a sample of size n. The estimator has mean

$$E(\hat{\theta}_W) = \theta, \tag{6.2}$$

and variance

$$\text{Var}(\hat{\theta}_W) = \frac{\theta}{a_n} + \frac{2\theta^2}{a_n^2 \rho^2} \int_0^\rho (\rho - x) f_n(x) \, dx \tag{6.3}$$

(see also equation (5.25)). If $\rho = 0$, then the integral in equation (6.3) is replaced by $b_n = \frac{1}{2} \sum_{j=1}^{n-1} 1/j^2$. The integral is always less than b_n. Watterson's estimator is a moment estimator derived from the formula $E(S_n) = \theta a_n$ by equating the mean $E(S_n)$ with the observed number of segregating sites.

A few remarks:

1. $\hat{\theta}_W$ is unbiased and the variance decreases with increasing ρ. The latter is expected as recombination breaks up linkage and reduces correlation between sites.

2. The variance of $\hat{\theta}_W$ eventually goes to zero as n goes to infinity, though slowly at rate $1/a_n$. Thus, $\hat{\theta}_W$ becomes more accurate with increasing n.

3. Population structure and changes in effective population size over time biases $\hat{\theta}_W$. Under exponential growth, $\hat{\theta}_W$ is downwards biased. Under migration, $\hat{\theta}_W$ is upwards biased, because the MRCA tends to be pushed further back in time than in a homogeneous population.

6.2.2   Tajima's estimator

Another commonly reported summary statistic is Tajima's estimator, $\hat{\pi}$ (see also Section 2.5.4). It plays an important role because of its conncection to Tajima's D (also introduced in Section 2.5.4). To compute $\hat{\pi}$ one must first compute all pairwise differences, π_{ij}, the number of sites that differ between sequence i and j. Clearly $\pi_{ij} = \pi_{ji}$. The set of pairwise differences are themselves a summary of the data; in that the sample configuration cannot in general be reconstructed uniquely from the π_{ij}s. Consider,

1.	0 0 0 0 0 0 0 0
2.	0 0 0 0 1 1 1 1
3.	0 0 1 1 0 0 1 1
4a.	0 1 0 1 0 1 0 1
4b.	0 0 1 1 1 1 0 0.

The sample consisting of sequence 1–3 and 4a has $\pi_{ij} = 4$ for all i and j. Similarly, the sample consisting of 1–3 and 4b has $\pi_{ij} = 4$. Thus, the sample configuration is not determined by the pairwise differences. In fact, in this example the π_{ij}s leave the impression of a star-shaped topology because all distances are equal. However, if the sample is compatible with a tree, then the sequences can be recovered apart from the order of the sites.

The statistic $\hat{\pi}$ is defined as the average of the π_{ij}s:

$$\hat{\pi} = \frac{2}{n(n-1)} \sum_{i<j} \pi_{ij}. \tag{6.4}$$

Since there are $n(n-1)/2$ pairs of sequences the sum in equation (6.4) is divided by $n(n-1)/2$ to obtain $\hat{\pi}$. In the basic coalescent the mean of π_{ij} is θ, because π_{ij} is the number of segregating sites in a sample of size two. As a consequence $E(\pi_{ij}) = \theta$, which in turn implies $E(\hat{\pi}) = \theta$ and that $\hat{\pi}$ is an unbiased estimator of θ. $\hat{\pi}$ is affected by population structure and changes in effective population size similarly to $\hat{\theta}_W$ and for the same reasons, though not to the same degree.

Under an arbitrary model

$$E(\hat{\theta}_W) = \frac{\theta E(L_n)}{2a_n}, \tag{6.5}$$

and

$$E(\hat{\pi}) = \theta E(T_2), \tag{6.6}$$

where L_n is the length of a tree relating n sequences and T_2 the TMRCA of two sequences. The direction of the bias of $\hat{\theta}_W$ and $\hat{\pi}$ is thus determined by $E(L_n)/2a_n$ and $E(T_2)$; the variables are upwards/downwards biased according to whether the two quantities are larger/smaller than one.

The variance of $\hat{\pi}$ is also known. It has a complicated integral representation, similarly to equation (6.3). For $\rho = 0$ the expression for the variance simplifies to

$$\mathrm{Var}(\hat{\pi}) = \frac{(n+1)\theta}{3(n-1)} + \frac{2(n^2+n+3)\theta^2}{9n(n-1)}. \tag{6.7}$$

As n becomes large, $\mathrm{Var}(\hat{\pi}) \to \theta/3 + 2\theta^2/9$ and $\hat{\pi}$ is not consistent in n (this is also true for $\rho > 0$). Intuitively, it makes sense: π_{ij} and $\pi_{i'j'}$ are highly correlated, as illustrated in Figure 6.3.

The mutation labelled 'a' in Figure 6.3 is counted four times, in $\pi_{13}, \pi_{14}, \pi_{23},$ and π_{24}, and 'b' is counted three times, in $\pi_{13}, \pi_{23},$ and π_{34}. (Note that this is true irrespective of whether the sequences are recombining or not.) Thus, π_{13} and π_{24} have an overlap of one mutation, and π_{13} and π_{23} an overlap of two. In general, mutations that are higher up in the tree, that is, nearer the root, are counted more times than those that are found nearer the tips.

Tajima's estimator is also a moment estimator obtained by equating $E(\hat{\pi})$ with the observed value of $\hat{\pi}$. As we will see in the next section Fu has taken full utility of moment estimators.

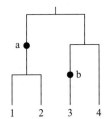

Figure 6.3 The mutation 'a' is counted in four pairwise differences, the mutation 'b' is counted in three.

6.2.3   Fu's two estimators

Neither $\hat{\theta}_W$ nor $\hat{\pi}$ use all information in the sample. Fu (1994) proposed two moment-based estimators to improve on $\hat{\theta}_W$ and $\hat{\pi}$. Both are fast to compute. He combined the requirement for unbiasedness with the require-ment that the variance of the estimator should be as small as possible under some further constraints.

Consider a series of statistics, $X_1, X_2, \ldots, X_K$, for example, the X_ks could be $\pi_{12}, \pi_{13}, \ldots, \pi_{n-1,n}$ and $K = n(n-1)/2$. Fu looked at unbiased linear combinations of the statistics, such as $2\pi_{12} + 3\pi_{13} - 4\pi_{23}$ or $\hat{\pi}$, both with mean θ. Each of these linear combinations is an estimator of θ. Among all the estimators of θ that are defined in this way one has particular interest: the linear combination, $\hat{\theta}$, with the least possible variance. For given vari-ables $X_1, X_2, \ldots, X_K$, an explicit expression for $\hat{\theta}$ can be found. In general, however, it turns out that in order to calculate $\hat{\theta}$ one needs to know θ itself. This is of course unfortunate because θ is the parameter we are trying to estimate and is thus unknown: To compute $\hat{\theta}$, θ is required; if θ is known, then we have little interest in estimating it. Fu resolved the circularity in the following way. The first step is to write $\hat{\theta}$ in terms of X_k and coefficients $a_k(\theta)$, that possibly depend on θ. These coefficients depend on the first two moments of X_k:

$$\hat{\theta} = \sum_{k=1}^{K} a_k(\theta) X_k. \tag{6.8}$$

Fu showed that the coefficients $a_k(\theta)$, $k = 1, \ldots, K$, can always be found, explicitly or by numerical calculations. The limiting factor is of course θ in $a_k(\theta)$ because that is the parameter we wish to estimate by $\hat{\theta}$ and hence it cannot be used in calculation of $\hat{\theta}$. Fu circumvented the problem defining an approximation of $\hat{\theta}$ recursively. It is done in the following way. Let

$$\hat{\theta}_0 = \sum_{k=1}^{K} a_k(\theta_0) X_k, \tag{6.9}$$

where θ_0 has a chosen fixed value, for example, $\theta_0 = 1$ or $\theta_0 = \hat{\theta}_W$, and

$$\hat{\theta}_i = \sum_{k=1}^{K} a_k(\hat{\theta}_{i-1}) X_k, \tag{6.10}$$

for $i \geq 1$. Repeat the calculation of $\hat{\theta}_i$ for $i = 1, 2, \ldots$, using equation (6.10) until $\hat{\theta}_i$ does not change significantly. Then $\hat{\theta}_i \approx \hat{\theta}$.

Fu's method is fast and efficient. Intuitively it compares the observed X_k, $k = 1, \ldots, K$, to the means $E(X_k)$ while taking the covariance structure into account, that is, taking $\text{Var}(X_k)$ and $\text{Cov}(X_k, X_{k'})$ into account.

6.2.3.1 Watterson's estimator revisited

It is useful briefly to return to the two estimators previously introduced. If we let $X_1 = S_n$ (that is $K = 1$), then the only unbiased linear combination of S_n is $\hat{\theta}_W = S_n/a_n$, that is, $a_1(\theta) = a_n$, and consequently $\hat{\theta}_W$ is also the linear combination with the least variance. Thus, Fu's approach can be seen as an extension and improvement of Watterson's estimator.

6.2.3.2 Tajima's estimator revisited

Tajima's estimator is one possible linear combination of π_{ij}, $1 \leq i < j \leq n$, with expectation θ. It can be shown that $\hat{\pi}$ is also the linear combination of the π_{ij}s with the least possible variance. Here, $a_k(\theta) = 2/(n(n-1))$, which is also independent of θ.

6.2.3.3 The i-Mutation estimator

Under the basic coalescent model, the expected number of i-mutations, without distinguishing between i- and $(n - i)$-mutations, is (see equation (2.34))

$$E(\eta_i) = \theta \left(\frac{1}{i} + \frac{1}{n-i} \right) \frac{1}{1 + \delta_{i,n-i}}, \qquad (6.11)$$

where $\delta_{i,j} = 1$ if $i = j$ and $\delta_{i,j} = 0$ otherwise. Again this is true irrespective of whether or not the sequences are recombining. The variables $\eta_1, \eta_2, \ldots, \eta_{n-1}$ play the role of $X_1, X_2, \ldots, X_K$, with $K = n - 1$. Fu also required the variances and covariance of η_i (i.e. X_i), and these are only known for $\rho = 0$. The expressions are lengthy and not very informative for our purpose. Hence, they will not be reproduced here.

Fu defined an estimator, $\hat{\theta}_F$, based on η_i and equations (6.10) and (6.9). It turned out that $\hat{\theta}_F$ is only a slightly better estimator than Watterson's, unless n is large (>50). However, when this condition is fulfilled, $\hat{\theta}_F$ tends to have a substantially lower variance than $\hat{\theta}_W$.

Both $\hat{\theta}_W$ and $\hat{\theta}_F$ are based on relating observed quantities to quantities averaged over all possible trees. Features in the data that point to a specific tree (or topology) are thus not used. However, when n becomes large then the number of mutations found in i copies is approximately Poisson with parameter θ/i, irrespective of what the topology is. This is a distinctive feature of the basic coalescent process, and $\hat{\theta}_F$ becomes effectively estimated from a series of independent Poisson variables. It explains why Fu's estimator has a better performance for large n.

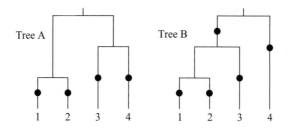

Figure 6.4   The two possible rooted topologies for four sequences. Branches that give rise to singleton mutations are marked with filled circles.

For small n the Poisson approximation does not hold. As an illustration, consider the two trees in Figure 6.4. The branching order in Tree A occurs with probability 1/3, whereas the branching order in Tree B occurs with probability 2/3. For Tree A the expected number of singletons, η_{1A}, is $2\theta/3$ and that of doublets, η_{2A}, is $7\theta/6$. Similarly for Tree B, the expected number of singletons, η_{1B}, is $5\theta/3$ and that of doublets, η_{2B}, is $\theta/6$. This agrees with equation (6.11): If the two expectations are weighed with the respective probabilities for the trees then $E(\eta_1) = \frac{1}{3}E(\eta_{1A}) + \frac{2}{3}E(\eta_{1B}) = \frac{1}{3}\frac{2}{3}\theta + \frac{2}{3}\frac{5}{3}\theta = \frac{4}{3}\theta$ and $E(\eta_2) = \frac{1}{3}E(\eta_{2A}) + \frac{2}{3}E(\eta_{2B}) = \frac{1}{2}\theta$. In Tree A seven times more doublets are expected than in Tree B. Neither $\hat{\theta}_W$ nor $\hat{\theta}_F$ are capable of taking this fact into account.

6.2.3.4   A phylogenetic estimator

Fu investigated another estimator, UPBLUE, using the same approach but without relying on η_i. The idea is to incorporate more information about the tree into the estimator. Ideally, one would take the coalescent topology with branch lengths as input, but this information is not available. Instead Fu suggested to use a tree, estimated from the sampled sequences. This can be done in numerous ways, the most important thing is to make sure the estimated tree respects a molecular clock, because that is an essential assumption of the coalescent process. Fu used an UPGMA tree. Then the branch lengths as well as the topology are the basic input, rather than η_i, in estimation of θ. This approach has the advantage of using information about the tree and the disadvantage of introducing a bias inherited in the tree estimation method. If the tree is wrong, then $\hat{\theta}$ might be even more wrong.

However, the variance of UPBLUE turned out to be significantly lower than both the variance of $\hat{\theta}_W$ and $\hat{\theta}_F$, in fact the variance is very close to the theoretical optimal. Indeed, this is interesting because it shows that information about the tree is crucial if accurate and reliable estimates are wanted. Fu's method is very fast and can easily compete computationally with more involved IS and MCMC methods.

Unfortunately, UPBLUE does not generalise to settings with recombination because it relies on estimating a tree prior to estimating θ. This is contrary to the other estimators that have been discussed until now. In fact, they tend to perform better with high scaled recombination rate than with low, simply because recombination reduces the variance in the number of mutations.

6.3 Estimators of ρ

Estimation of the scaled recombination rate is a harder problem, both statistically and computationally, than estimation of the scaled mutation rate. In the beginning of Section 6.2 we mentioned one reason why this is so; here we list two more reasons.

1. Under the infinite-site assumption all mutation events in a sample history can be listed whereas all recombination events cannot. A recombination event can only be inferred with certainty if all four gametes are present. Thus, there seems to be less information about the recombination rate than about the mutation rate in the sample.

2. The number of possible genealogical relationships between sequences subject to recombination is unlimited, whereas it is finite, though large, for non-recombining sequences.

The second reason introduces a computational limitation that might be relaxed in the future with faster computers. The first reason is statistical and cannot be improved upon; only the methods by which data is analysed can be improved. Most recombination events cannot be detected using the four-gamete test. This is true even if the scaled mutation rate is extremely high and the sample size is large. For example, if $n = 1000$ and mutations are dense on every branch, at most 69% of all recombination events are picked up by the four-gamete test. For smaller sample sizes and realistic mutation rates this number is much less than 69%.

In Chapter 5 the summary statistics T_M and H_M were discussed as indicators of recombination. Both T_M and H_M might overlook important information in the data: They merely provide lower bounds to the number of recombination events the sample must have experienced in its history. For example, consider the following extreme case, where T_M and H_M fail to detect recombination, though it still can be argued that the data contains evidence in favour of recombination. Assume two non-recombining loci with rate ρ between them. One could think of the two loci as being separated by a recombination hotspot.

The likelihood as a function of ρ, θ_1, and θ_2 (θ_i is the scaled mutation rate in locus i), is

$$L(\theta_1, \theta_2, \rho) = P_{\theta_1, \theta_2, \rho}(S_{1,n} = k_1, S_{2,n} = k_2), \tag{6.12}$$

where k_1 and k_2 are the number of segregating sites in the two loci, respectively. Subscripts are used to indicate dependence on θ_i and ρ. For $n = 2$, the likelihood is easy to calculate by hand in two extreme cases, complete linkage ($\rho = 0$) and no linkage at all ($\rho = \infty$), for given values of k_1, k_2, θ_1, and θ_2:

$$L(\theta_1, \theta_2, 0) = \frac{(k_1 + k_2)!}{k_1! k_2!} \frac{\theta_1^{k_1} \theta_2^{k_2}}{(1 + \theta_1 + \theta_2)^{1+k_1+k_2}}, \tag{6.13}$$

and

$$L(\theta_1, \theta_2, \infty) = \frac{\theta_1^{k_1} \theta_2^{k_2}}{(1 + \theta_1)^{1+k_1}(1 + \theta_2)^{1+k_2}}. \tag{6.14}$$

Assume the mutation rates (u_1 and u_2) in the two genes are identical and that the genes are roughly of the same length, then $\theta_1 = \theta_2$. If $k_1 = 1$ and $k_2 = 7$,

$$L(4, 4, 0) = 0.0014 \quad \text{and} \quad L(4, 4, \infty) = 0.0067, \tag{6.15}$$

where $\hat{\theta}_L = 4$ is the maximum likelihood estimator of $\theta = \theta_1 = \theta_2$ for both $\rho = 0$ and $\rho = \infty$. In other words the likelihood supports two unlinked loci more than two completely linked loci. Intuitively this makes sense because in the first locus only one mutation is observed and a short tree is expected, whereas in the second locus seven mutations are observed and a deep tree is expected. Since $L(4, 4, \infty) > L(4, 4, 0)$, $\hat{\rho} > 0$ even though the data passes the four-gamete test and has $T_M = H_M = 0$. Thus, it can be concluded that recombination can be inferred even in the absence of incompatibilities. In contrast, if $k_1 = k_2 = 4$ then the likelihood supports two complete linked loci.

Some further comments are appropriate:

1. In the absence of any information about θ_1 and θ_2 it would be natural to expect $\theta_1 \neq \theta_2$, because seven times as many mutations are observed in locus 2 as in locus 1. The same conclusion as above cannot be reached if θ_1 and θ_2 are estimated separately. In that case, $L(1, 7, 0) = 0.017$ and $L(1, 7, \infty) = 0.012$, where $\hat{\theta}_{L,1} = 1$ and $\hat{\theta}_{L,2} = 7$ for both $\rho = 0$ and $\rho = \infty$.

2. If we have firm belief in $u_1 = u_2$, but know the genes have unequal length, the calculation could be carried out given that $\theta_1 = c\theta_2$ for some known constant (c is the relative length of locus 1 to locus 2).

3. The example is silly in the sense that no one would dream of inferring recombination from such a tiny data set. However, it illustrates the points that (a) information about recombination is not limited to incompatibilities, and (b) the model assumptions can play a major role in the conclusions that can be drawn from the analysis.

Recombination is difficult to take into account in an analysis and it is very tempting to ignore it or assume that the effects of recombination are minor and will not affect the statistical conclusions. This might of course have very unfortunate consequences. It is well known that estimation of trees and times of events become biased if recombination is ignored when actually present (e.g. see Schierup and Hein 2000a, b). Generally more mutation events are required to explain the data if recombination is assumed absent, because the infinite sites assumption might not hold. Four sequences with two incompatible sites require at least three mutation events if recombination is not allowed, while two are sufficient if recombination is allowed.

6.3.1 Estimators based on summary statistics

As mentioned earlier summary statistics are appealing because data becomes compressed into a form that often allows for simple data analysis. S_n and T_M are two frequently encountered summary statistics. Intuitively, S_n summarises the available information about θ, whereas T_M summarises information about ρ. From S_n it is easy to obtain an estimate of θ that applies even in the presence of recombination. In contrast it is not possible to obtain an estimate of ρ from T_M alone. For example, a value of T_M might be explained by many different combinations of θ and ρ. Which one to apply? In contrast, it is possible to estimate θ and ρ jointly from S_n and T_M. In this section various estimators based on summary statistics are reviewed.

6.3.1.1 *Wakeley's estimator*

Wakeley (1997) suggested an estimator based on moments of $\hat{\pi}$ (Section 6.2.2) and the empirical variance of the π_{ij}s,

$$\hat{\pi}_2 = \frac{2}{n(n-1)} \sum_{i<j} (\pi_{ij} - \hat{\pi})^2. \tag{6.16}$$

Note that in the definition of $\hat{\pi}_2$ the observed $\hat{\pi}$ is used, rather than the expectation of $\hat{\pi}$, $E(\hat{\pi}) = \theta$. Thus $\hat{\pi}_2$ can be found without any knowledge of θ. The empirical variance $\hat{\pi}_2$ has mean

$$E(\hat{\pi}_2) = \frac{2(n-2)}{3(n-1)}\theta + g_n(\rho)\theta^2, \tag{6.17}$$

where $g_n(\rho)$ is a complicated function of ρ, whose form is fully known.

Now, if reasonable estimates of θ and θ^2 are available, an estimate of ρ can be found be solving the equation (6.17) with $E(\hat{\pi}_2)$ replaced by the observed value of $\hat{\pi}_2$. Wakeley suggested to use the unbiased estimate $\hat{\pi}$ of θ and the unbiased estimate

$$\hat{\theta^2} = \frac{\hat{\pi}^2 - [(n+1)/3(n-1)]\hat{\pi}}{f_n(\rho)+1}$$

of θ^2, where $f_n(\rho)$ is as defined in Section 5.7. The latter he found by considering the expression for the variance of $\hat{\pi}$.

Generally, Wakeley's estimator has large variance, partly because $\hat{\pi}$ has large variance. This is a feature it shares with other moment estimators of ρ due to lack of sensitivity. The expectation of $\hat{\pi}_2$ does not strongly depend on ρ and it becomes difficult to distinguish between different values of ρ. Including uncertainty about θ into the equation makes the task of estimation of ρ from moments even more difficult.

6.3.1.2 Likelihood and summary statistics

Wall (2000) suggested an inferential procedure for ρ based on the likelihood of (S_n, K_n, T_M), where K_n is the number of haplotypes in the sample. The likelihood is easily evaluated through simulations, because S_n, K_n, and T_M attain only integer values and thus a relatively small number of simulations are required to estimate the likelihood accurately. If the data set is large, simulations can be speeded up by fixing θ, for example, letting $\theta = \hat{\theta}_W$, or by fixing the number of segregating sites. The latter is done by tossing exactly S_n mutations onto every simulated genealogy; thereby only ρ is estimated from the data and no assumptions about θ are made. This procedure is often referred to as 'conditioning on S_n'. The terminology is, however, incorrect from a probabilistic point of view. The distribution of the genealogy conditioned on S_n depends on both S_n and θ, whereas the distribution of the genealogy with precisely S_n mutations imposed does not depend on S_n nor on θ, by construction. For example, for $n = 2$ the distribution of T_2 *given* $S_2 = k$ is Gamma distributed, $\Gamma(k+1, \theta+1)$ (see Section 3.8.2), whereas the distribution of T_2 with S_n mutations imposed is exponentially distributed, $\text{Exp}(1)$, whatever k and θ are.

A likelihood approach based on summary statistics is attractive for at least the following reasons: (1) it is likelihood based; (2) it is easy to implement;

and (3) problems with accuracy and correctness of the simulation procedure (e.g. MCMC) are generally circumvented.

6.3.2 Pseudo-likelihood estimators

Pseudo-likelihood methods approximate the likelihood with a quantity that is easier or faster to compute than the likelihood itself. We will only consider one example of a pseudo-likelihood method, but others exist, for example, Fearnhead and Donnelly (2001).

6.3.2.1 *Hudson's pairwise estimator*

Hudson (2001*b*) considered all pairs of segregating sites, ignoring all non-polymorphic sites in the sample. For each pair (i, j) it is straightforward to calculate the probability of obtaining the observed pattern of gametes given both sites are segregating. By considering only two sites, the scaled mutation rate θ is almost zero. As a consequence, the aforementioned probability depends only on the number of each gamete type and the scaled recombination rate between the sites. Let $\mathbf{n}_{ij}$ denote the vector of gamete counts for sites i and j (possible gametes are 00, 01, 10, and 11), and let ρ_{ij} be the scaled recombination rate, which is assumed to be proportional to in the number of nucleotides between i and j. That is $\rho_{ij} = \rho L_{ij}/L$, where L_{ij} is the number of nucleotides between i and j, and L the total number of nucleotides in a sequence.

The probability of obtaining the counts $\mathbf{n}_{ij}$ given that both i and j are polymorphic is

$$\frac{P_{\rho_{ij}}(\mathbf{n}_{ij})}{P_{\rho_{ij}}(i \text{ and } j \text{ are polymorphic})}. \tag{6.18}$$

Hudson proposed to estimate ρ from the pseudo-likelihood function

$$L_{\text{pseudo}}(\rho) = \prod_{(i,j)} \frac{P_{\rho_{ij}}(\mathbf{n}_{ij})}{P_{\rho_{ij}}(i \text{ and } j \text{ are polymorphic})}, \tag{6.19}$$

where the product is over all pairs of polymorphic sites. It depends on ρ only, because ρ_{ij} is assumed proportional to sequence length. Equation (6.18) is a pseudo-likelihood function, rather than a likelihood function, because

1. Only the likelihood of pairs of segregating sites are considered.
2. Pairs are treated as independent of each other (which they are not).
3. The likelihood of a pair is conditioned on the pair being segregating in both loci.

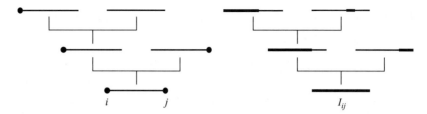

Figure 6.5   The history of two sites has generally fewer recombination events than the history of a whole segment of sites. The second recombination event does not contribute to the history of i and j.

Computation of $L_{pseudo}(\rho)$ is much less demanding than computation of the full likelihood of the sample. In principle, both quantities require summation over all possible genealogies. The genealogy of a pair of sites (i, j) (with no information about segregation pattern) is simpler than the genealogy of all sites from i to j. Let this segment of sites be I_{ij}. The genealogy of the pair (i, j) has fewer recombination events than the genealogy of I_{ij}, as illustrated in Figure 6.5. Only recombination events that happen in sequences for which both i and j are ancestral to the sample affect the genealogy of (i, j). In contrast, all events that happen in ancestral material between i and j affect the genealogy of I_{ij}. In the left part of the figure the second recombination event is not part of the history of i and j, whereas it is part of the history of I_{ij}.

This simplification speeds up computation greatly. For example, if $n = 2$, then the expected number, E_{ij}, of recombination events in the history of two sites i and j is

$$E_{ij} = \frac{2\rho_{ij}(2\rho_{ij} + 9)}{\rho_{ij}^2 + 13\rho_{ij} + 18}, \tag{6.20}$$

whereas the expected number for the whole segment I_{ij} is

$$E(R_2) = \rho_{ij}. \tag{6.21}$$

Thus, the ratio of the two quantities is

$$\frac{E_{ij}}{E(R_2)} = \frac{2(2\rho_{ij} + 9)}{\rho_{ij}^2 + 13\rho_{ij} + 18}, \tag{6.22}$$

which, for example, is $22/32 \approx 0.69$ for $\rho_{ij} = 1$, and $29/124 \approx 0.23$ for $\rho_{ij} = 10$.

6.4 Monte Carlo methods

The principle behind Monte Carlo methods is the following. Say one seeks the mean of a random variable X, or of any function $g(X)$ of X, for example, the expected outcome of a throw of a dice. Then, throw the dice several times and calculate the average overall values obtained. This does not give the true mean but if the dice is thrown many times the average comes close to the true mean. For example, the first five throws might yield the values: 2, 6, 2, 5, 1, with an average $(2+6+2+5+1)/5 = 16/5 = 3.2$, while the true mean is $(1+2+3+4+5+6)/6 = 3.5$. Formally, the Monte Carlo principle states

$$E(g(X)) \approx \frac{1}{M} \sum_{j=1}^{M} g(x_j), \tag{6.23}$$

where M is the number of times the variable X is simulated (thrown) and the $x_1, \ldots, x_M$ are the outcomes of the M simulations. The mean is said to be obtained by Monte Carlo integration of $g(X)$ with respect to X and the distribution of X is called the sample distribution or the proposal distribution. This approach guarantees that the right-hand side of (6.23) is an unbiased estimator of $E(g(X))$, but not necessarily that it has a small error. In the example with the dice the error was $3.5 - 3.2 = 0.3$. The error tends to disappear as M becomes large. One quantity that reflects the error is

$$\frac{\text{Var}(g(X))}{M}, \tag{6.24}$$

where $\text{Var}(g(X))$ is the variance of $g(X)$. Often the variance is not known and must also be estimated from $x_1, \ldots, x_M$.

It is illustrative to consider another example, one that at the same time is informative about the Monte Carlo method and is still easy to handle. So, again we turn to $P(S_n = k)$ and this time seek the value of $P(S_n = k)$ from Monte Carlo integration. The exact value of the probability can be found in many ways, one is to apply the recursion (2.28). Thus, the accuracy of Monte Carlo simulations can easily be checked against the true value.

The initial step is to write the probability $P(S_n = k)$ as the mean of a random variable. This can be done in numerous ways. For example,

$$P(S_n = k) = E(1_{\{S_n=k\}}(S_n)) \approx \frac{1}{M} \sum_{j=1}^{M} 1_{\{S_n=k\}}(k_j), \tag{6.25}$$

where $1_A(X)$ is an indicator function that is one if X is in A and zero otherwise, and $k_1,\ldots,k_M$ are the number of segregating sites in M simulations of n genes. The idea is to simulate genealogies of n genes, add mutations and count the number of times a genealogy has exactly k mutations. Unfortunately, many simulated genealogies will not contribute to the sum at all: If $k_j \neq k$, it counts as zero in the sum. This is reflected in the error measure, the variance of $1_{\{S_n=k\}}$, to which we will return later.

The probability $P(S_n = k)$ can be given in the form of an integral,

$$P(S_n = k) = \frac{(n-1)\,\theta^k}{k!\,2^{k+1}} \int_0^\infty x^k\, e^{-\frac{(\theta+1)x}{2}} \left(1 - e^{-\frac{x}{2}}\right)^{n-2} dx \qquad (6.26)$$

(Tavaré 1984). If the terms under the integral are rearranged $P(S_n = k)$ can be characterised as an expectation of a random variable in two other ways. The first is given by

$$P(S_n = k) = \frac{(n-1)}{\theta} E\left[e^{-\frac{X}{2}}\left(1 - e^{-\frac{X}{2}}\right)^{n-2}\right], \qquad (6.27)$$

where the proposal X is gamma distributed with parameters $k+1$ and $\theta/2$. The second is given by

$$P(S_n = k) = \frac{\theta^k}{k!\,2^k} E\left[L_n^k\, e^{-\frac{\theta L_n}{2}}\right], \qquad (6.28)$$

where the proposal L_n is the sum of all branches in the coalescent tree. Intuitively equations (6.27) and (6.28) form a better basis for Monte Carlo simulation of $P(S_n = k)$ than equation (6.25) because all simulated values of X and L_n, respectively, contribute to the probability. The three approaches are compared in Table 6.1.

The probability $P(S_n = k)$ can be evaluated in many other ways depending on how the integral in (6.26) is decomposed, $\int g(x)f(x)\,dx = E(g(X))$, such that $f(x)$ is the density of the proposal, X, and $g(x)$ a function of that variable. The efficiency of the Monte Carlo method is improved by choosing carefully the variable, X, under which simulations are performed. Choosing $f(x)$ (and $g(x)$) is called Importance Sampling (IS). In (6.25), most of the simulated values do not contribute to the sum at all, whereas in (6.27) and (6.28) all of them do. This reduces the burden of computation significantly, and the accuracy of the estimated probability. Table 6.1 shows the standard deviation of $g(X)$ used in three Monte Carlo simulation of $P(S_n = k)$.

Table 6.1   Calculation of $P(S_{10} = k)$, $n = 10$, using three different Monte Carlo approaches, (6.25), (6.27), and (6.28), for various values of k and θ^a

θ	k	$P(S_{10} = k)$	Standard Deviation		
			Equation (6.25)	Equation (6.27)	Equation (6.28)
	1	0.19	0.39	0.14	0.10
1	3	0.18	0.38	0.13	0.05
	5	0.08	0.27	0.11	0.05
	3	0.06	0.24	0.04	0.06
3	6	0.10	0.30	0.03	0.05
	9	0.08	0.27	0.04	0.04

^{a} The expected number of mutations is $\theta \sum_{j=1}^{9}(1/j)$ which is ≈ 2.8 for $\theta = 1$ and ≈ 8.5 for $\theta = 3$. A total of 10^5 simulations were performed for each entry.

As expected, (6.25) is poor, whereas both (6.27) and (6.28) perform better.

6.4.1   The likelihood curve

The usefulness of the Monte Carlo and IS methods becomes even more transparent when the aim is not just to evaluate $P(S_n = k)$ for a single value of θ but for a whole range of θ-values. For example, assume k mutations are observed in a sample of n genes and that the true value of θ is not known but sought estimated from the data, for example, using maximum likelihood. That is, we would like to calculate the likelihood $L(\theta) = P_\theta(S_n = k)$ for a large range of θ-values and single out the value that has the highest probability, that is, the maximum likelihood estimator of θ.

The IS scheme can immediately be invoked here. For example, from equation (6.28): Simulate $y_1, \ldots, y_M$ from L_n and calculate the empirical average

$$L(\theta) = P_\theta(S_n = k) \approx \frac{\theta^k}{M\, k!2^k} \sum_{j=1}^{M} y_j^k\, e^{-\frac{\theta y_j}{2}}. \tag{6.29}$$

Note that only one set of simulations is performed and that these simulations are used to calculate the likelihood for all θ.

Alternatively one can extend the idea used in (6.28). Consider some fixed θ_0. Then for any θ, equation (6.26) can be written as

$$P_\theta(S_n = k) = \frac{(n-1)\,\theta^k}{k!\,2^{k+1}} \int_0^\infty x^k\, e^{-\frac{\theta_0 x}{2}} e^{-\frac{(\theta - \theta_0 + 1)x}{2}} \left(1 - e^{-\frac{x}{2}}\right)^{n-2} dx. \tag{6.30}$$

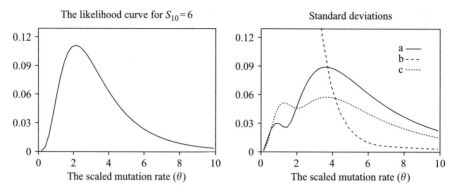

Figure 6.6   The likelihood curve as a function of θ for $P_\theta(S_{10} = 6)$ and the standard deviations obtained with different approaches. (a) is based on equation (6.29), (b) on equation (6.31) with $\theta_0 = 5$, and (c) on equation (6.31) with $\theta_0 = 2.2$. 10^6 simulation were performed to obtain the two curves, apart from the curve for $\theta_0 = 5$ where 10^8 simulations were performed to obtain an accurate estimate of the standard deviation.

Using the Monte Carlo technique the integral can be approximated by

$$P_\theta(S_n = k) \approx \frac{(n-1)\theta^k}{\theta_0^{k+1}} \sum_{j=1}^{M} e^{-\frac{(\theta-\theta_0+1)x_j}{2}} \left(1 - e^{-\frac{x_j}{2}}\right)^{n-2}, \qquad (6.31)$$

where $x_1, \ldots, x_M$ are M values obtained from the proposal distribution $\Gamma(k + 1, \theta_0)$ which is dependent on the choice of θ_0. In the literature θ_0 is called the 'driving value'. An appropriate choice of θ_0 could be a simple estimator of θ, for example, Watterson's estimator. Figure 6.6 shows the likelihood curve, $L(\theta) = P_\theta(S_n = k)$ and standard deviations of $g(X)$ obtained from Monte Carlo simulations.

The example we have walked through is extremely simple. It is possible, and fast, to compute the exact value of $P_\theta(S_n = k)$ for a large range of θ using recursion (2.28). The alternative approaches, based on Monte Carlo integration and IS, are in the example also very fast and very reliable.

In the first part of Figure 6.6 the exact likelihood curve, $L(\theta) = P_\theta(S_{10} = 6)$, is shown. It attains it maximum for $\theta \approx 2.2$, that is, $\hat{\theta} \approx 2.2$. In the second part the standard deviations of $g(X)$ in (6.29) and (6.31) are shown. In one case Watterson's estimator was chosen as driving value, $\theta_0 = 2.2$, in the other case $\theta_0 = 5$ was chosen. For $\theta_0 = 5$ the standard deviation rises to about 2.5 around the maximum $\hat{\theta} = 2.2$. Note that $\hat{\theta}$ is remarkably close to Watterson's estimator.

6.4.2 Monte Carlo integration and the coalescent

Generally, there is not an explicit expression for the full likelihood, $L(\theta)$, of a sample under a coalescent model. It can be written in the form

$$L(\theta) = P_\theta(D) = \int_{\mathcal{H}} P_\theta(D|\mathcal{H})P_\theta(\mathcal{H})\,d\mathcal{H}, \qquad (6.32)$$

where the integral is over all possible histories $\mathcal{H}$ of the data, D. (θ is used as an example parameter; any other parameter(s) could essentially be substituted.) Equation (6.32) also suggests that the likelihood can be approximated using Monte Carlo techniques, as discussed and illustrated in the previous section.

To do Monte Carlo integration of equation (6.32) we first need to define $\mathcal{H}$ precisely. Some methods proposed in the literature include the ancestral states of sequences and mutation events, in addition to coalescent events, in the definition of $\mathcal{H}$. Other methods define $\mathcal{H}$ by a series of coalescent events. These definitions are somewhat related to the two algorithms presented in Chapter 2 to generate a sample under the coalescent model: One can generate coalescent and mutation events at the same time, or generate the genealogy first and superimpose mutation events onto the genealogy. In Figure 6.7 the history of the left tree is three coalescent events, of the right tree it is three coalescent events and two mutation events.

For now, let $\mathcal{H}$ be shorthand for a set of coalescent times and a branching order, for example, first sequence 3 and 5 coalesce, then the ancestor of 3 and 5 coalesces with 8, etc. That is $\mathcal{H}$ can be written $\mathcal{H} = (T, W)$, where T is a coalescent topology (the branching order) and W the times between coalescent events. We will refer to the space of all possible histories $\mathcal{H}$ as *tree space*. Thus, $P_\theta(D|\mathcal{H})$ is the probability of obtaining the data D, given the genealogical history $\mathcal{H}$. For any given coalescent topology and set of coalescent times $P_\theta(D|\mathcal{H})$ is easily evaluated. Integrating $P_\theta(D|\mathcal{H})$ over W is possible, but not easy. However, because there are $n!(n-1)!/2^{n-1}$ different coalescent topologies it is practically impossible to sum up all these, even for modest sample sizes. For example, the number of coalescent

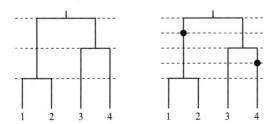

Figure 6.7 In the left figure only coalescent events define the history. In the right figure both coalescent events and mutation events define the history.

topologies reaches $2.57 \cdot 10^9$ for $n = 10$. Thus, it is not possible to evaluate the likelihood over the whole tree space. Note that if recombination is allowed the elements in tree space would no longer be trees but graphs.

As a first impulse one could choose histories $\mathcal{H}_i = (\mathcal{T}_i, \mathcal{W}_i)$, $i = 1, \ldots, M$, randomly according to the coalescent model a large number of times, M, and approximate $L(\theta)$ with the average of $P_\theta(D \mid \mathcal{H}_i)$;

$$L(\theta) \approx \frac{1}{M} \sum_{i=i}^{M} P_\theta(D \mid \mathcal{H}_i). \tag{6.33}$$

This is a straightforward Monte Carlo approach that, however, is also not efficient: Most of the coalescent topologies, $\mathcal{T}_i$, will not be compatible with D (except in a few special cases); thus, most of the simulations are wasted because these histories do not contribute anything to the likelihood, that is, $P_\theta(D|\mathcal{H}_i) = 0$. For example, with four sequences there are eighteen possible coalescent topologies of the tree (Figure 6.8): Tree A gives rise to six different labelled coalescent topologies, whereas Tree B gives rise to twelve. If there is one segregating site in the data separating genes 1 and 2 from genes 3 and 4, then there are six compatible coalescent topologies, of which two are shown in Figure 6.8. Since all coalescent topologies are equally likely, only 1/3 (six out of eighteen) of all simulations count in the likelihood. This fraction becomes decreasingly smaller with the number of genes.

6.4.2.1 *Importance sampling (IS)*

Importance sampling aims to reduce the variance of the estimated probability and to reduce the number of simulations that contribute little in the likelihood. Instead of choosing histories from the distribution $P_\theta(\mathcal{H})$, one would like to sample histories from a proposal distribution, $Q(\mathcal{H})$, that provides better support for the observed data, D. To accomplish this,

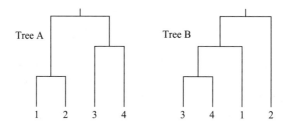

Figure 6.8 There are two possible topologies relating four sequences. A can be labelled in six ways, B in twelve. In A, two of these are compatible with a split that separates (1, 2) from (3, 4); in B, four are compatible with this split.

reformulate equation (6.32),

$$L(\theta) = \int_{\mathcal{H}} P_\theta(D \mid \mathcal{H}) \frac{P_\theta(\mathcal{H})}{Q(\mathcal{H})} Q(\mathcal{H}) \, d\mathcal{H}, \qquad (6.34)$$

where the proposal $Q(\mathcal{H})$ is a distribution on tree space; $Q(\mathcal{H})$ is not necessarily a coalescent process (in fact, neither in Griffiths and Tavaré's (1994) work nor in Stephens and Donnelly's (2000) is $Q(\mathcal{H})$ the standard coalescent process). Now, the likelihood of the data can be approximated by

$$L(\theta) \approx \frac{1}{M} \sum_{i=1}^{M} P_\theta(D \mid \mathcal{H}_i) \frac{P_\theta(\mathcal{H}_i)}{Q(\mathcal{H}_i)}, \qquad (6.35)$$

which reduces to equation (6.33) if $Q(\mathcal{H}) = P_\theta(\mathcal{H})$. Note that (6.34) allows us to approximate the likelihood curve using a single choice of $Q(\mathcal{H})$. The histories simulated under $Q(\mathcal{H})$ can be used to calculate the likelihood for any choice of θ, thus also for the whole likelihood curve.

Ideally, one would like to sample from $Q(\mathcal{H}) = \tilde{Q}(\mathcal{H})$, where

$$\tilde{Q}(\mathcal{H}) = P_\theta(\mathcal{H} \mid D) = P_\theta(D \mid \mathcal{H}) \frac{P_\theta(\mathcal{H})}{P_\theta(D)} = P_\theta(D \mid \mathcal{H}) \frac{P_\theta(\mathcal{H})}{L(\theta)}, \qquad (6.36)$$

because in that case the approximation becomes exact,

$$\frac{1}{M} \sum_{i=1}^{M} P_\theta(D \mid \mathcal{H}_i) \frac{P_\theta(\mathcal{H}_i)}{\tilde{Q}(\mathcal{H}_i)} = \frac{1}{M} \sum_{i=1}^{M} L(\theta) = L(\theta). \qquad (6.37)$$

In fact, only $M = 1$ simulation is necessary. However, knowledge of $\tilde{Q}(\mathcal{H})$, for any $\mathcal{H}$, is equivalent to knowledge of $L(\theta)$, and therefore not a feasible approach. A proposal distribution between the two extremes $P_\theta(\mathcal{H})$ and $\tilde{Q}(\mathcal{H})$ is required. Griffiths and Tavaré (1994) came up with one choice of proposal function that always provided histories compatible with D. Stephens and Donnelly (2000) came up with a different choice that reduced the burden of computation significantly and the error of the approximation, compared to the proposal suggested by Griffiths and Tavaré (1994).

Griffiths and Tavaré (1994) used the recursion discussed in Section 2.4.2 to define their proposal distribution. Consider the example in Figure 2.13. A history (proposal) $\mathcal{H}$ is defined as a path through the diagram, from the initial configuration until there is only one ancestral gene. In consequence a history also includes mutation events. $\mathcal{H}$ has probability defined by the product of weights attached to the edges that belong to $\mathcal{H}$, for example, if

$\mathcal{H}$ follows the right edge then the first term in the probability of $\mathcal{H}$ is

$$\frac{1}{4+\theta_0}\left(\frac{5+2\theta_0}{5(4+\theta_0)}\right)^{-1} = \frac{5}{5+2\theta_0}, \tag{6.38}$$

where θ_0 is the driving value. The history $\mathcal{H}'$ defined by the rightmost path up through the diagram has probability

$$Q(\mathcal{H}') = \frac{5}{(5+2\theta_0)}\frac{1}{2}\frac{2}{(2+\theta_0)} = \frac{5}{(5+2\theta_0)(2+\theta_0)}. \tag{6.39}$$

The last five terms in $Q(\mathcal{H}')$ turn out to be one because there is only one edge attached to the corresponding configurations. In equation (6.35),

$$P_\theta(D\,|\,\mathcal{H}') = 1, \tag{6.40}$$

because by construction all histories are consistent with the data, and

$$P_\theta(\mathcal{H}') = \frac{4}{(4+\theta)}\frac{1}{4}\frac{\theta}{(3+\theta)}\frac{1}{2}\frac{3}{(3+\theta)}\frac{2}{3}\frac{\theta}{(2+\theta)}\frac{1}{3}\frac{2}{(2+\theta)}\frac{1}{2}\frac{\theta}{(1+\theta)}$$
$$\times \frac{1}{2}\frac{\theta}{(1+\theta)}\frac{1}{(1+\theta)}$$
$$= \frac{\theta^4}{6(4+\theta)(3+\theta)^2(2+\theta)^2(1+\theta)^3} \tag{6.41}$$

is the product of coalescent probabilities for the events defining the history: first a coalescent event, then a mutation, a coalescent, a mutation, a coalescent, a mutation, a mutation, and finally a coalescent event. The factor in front of a fraction is the probability that a mutation happens in a given lineage(s) or that a coalescent event happens amongst certain pairs of genes. For example, $1/4$ is the probability that a mutation happens in the lineage that already has experienced a coalescent event. The history contributes

$$P_\theta(D\,|\,\mathcal{H}')\frac{P_\theta(\mathcal{H}')}{Q(\mathcal{H}')} = \frac{\theta^4(5+2\theta_0)(2+\theta_0)}{30(4+\theta)(3+\theta)^2(2+\theta)^2(1+\theta)^3} \tag{6.42}$$

to the Monte Carlo sum in equation (6.35).
   In the notation of Section 6.4

$$g(\mathcal{H}) = P_\theta(D\,|\,\mathcal{H})\frac{P_\theta(\mathcal{H})}{Q(\mathcal{H})}, \tag{6.43}$$

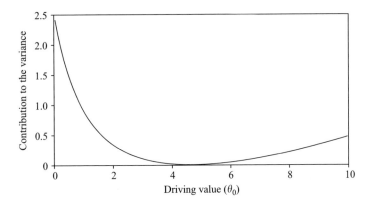

Figure 6.9   Shown is the contribution to the variance from the history $\mathcal{H}'$ using the scheme devised by Griffiths and Tavaré (1994). Here, $\theta = 2.12$ which is the maximum likelihood estimate. Small and large driving values, θ_0, produce large contributions to the variance. The weight is in units of 10^{-7}.

which has mean $L(\theta)$, by construction. The contribution from $\mathcal{H}'$ to the variance of $g(\mathcal{H})$ is

$$\left(g(\mathcal{H}') - L(\theta)\right)^2 Q(\mathcal{H}'), \qquad (6.44)$$

where $Q(\mathcal{H}')$ is the probability of obtaining $\mathcal{H}'$ and the first term is the squared deviance from the mean. As argued in Section 6.4 the variance is a measure of efficiency of the proposal distribution. The effect of chosing the driving value is clear from Figure 6.9: In this case it should neither be too small nor too large, where θ is the maximum likelihood estimate, $\hat{\theta}_L = 2.12$ (see Section 2.4.2) and the driving value θ_0 is varied. The least contribution is obtained for $\theta_0 \approx 4.7$, which is somewhat larger than the maximum likelihood estimate. Equation (6.42) depends quadratically on θ_0 while equation (6.39) is inversely proportional to θ_0. Thus when one of them is large the other is small and vice versa. To obtain a small contribution to the variance these two effects need to be balanced. In the end, of course, the average over all histories determines whether a driving value is appropriate or not.

Stephens and Donnelly (2000) improved on Griffiths and Tavaré's proposal distribution, by reducing the number of times histories that make small contributions to the likelihood are sampled.

6.4.3   Markov chain Monte Carlo

Kuhner et al. (1995, 1998) pursued a different approach. Their approach assumes a finite sites model, unlike Griffiths and Tavaré (1994) and Stephens and Donnelly (2000) who developed software both for the infinite sites model and finite sites model. One main difference between the two kinds of models is that all coalescent topologies are compatible with data from a finite sites model (though some might have very little probability of occurring) whereas only some coalescent topologies are compatible with data from an infinite sites model.

Kuhner et al. (1995, 1998) used IS to calculate the likelihood ratio which they wrote in the form

$$\frac{L(\theta)}{L(\theta_0)} = \int \frac{P_\theta(D\,|\,\mathcal{H})}{L(\theta_0)} P_\theta(\mathcal{H})\,d\mathcal{H} = \int \frac{P_\theta(D\,|\,\mathcal{H})\,P_\theta(\mathcal{H})}{P_{\theta_0}(D\,|\,\mathcal{H})\,P_{\theta_0}(\mathcal{H})}$$

$$\times\, P_{\theta_0}(\mathcal{H}|D)\,d\mathcal{H} \qquad (6.45)$$

(using Bayes' formula) and approximated by

$$\frac{L(\theta)}{L(\theta_0)} \approx \frac{1}{M} \sum_{i=1}^{M} \frac{P_\theta(D\,|\,\mathcal{H}_i)\,P_\theta(\mathcal{H}_i)}{P_{\theta_0}(D\,|\,\mathcal{H}_i)\,P_{\theta_0}(\mathcal{H}_i)}, \qquad (6.46)$$

that is, the IS function is $Q(\mathcal{H}) = P_{\theta_0}(\mathcal{H}\,|\,D)$. However, instead of independent sampling of histories from $P_{\theta_0}(\mathcal{H}\,|\,D)$, which in the previous section was found to be difficult, Kuhner et al. used the Metropolis–Hastings algorithm to construct a Markov Chain with distribution $P_{\theta_0}(\mathcal{H}\,|\,D)$. The algorithm guarantees that in the long run the simulated histories $\mathcal{H}_i$, $i = 1,\ldots,M$, follow the distribution $P_{\theta_0}(\mathcal{H}\,|\,D)$, at the cost that histories no longer are independent; the choice of $\mathcal{H}_{i+1}$ depends on $\mathcal{H}_i$. The benefit of this approach is that the Markov Chain tends to stay in areas of the tree space that support the data well before moving on to another area (see Figure 6.10), unlike independent sampling where sampling of one 'good' tree does not improve the chance of sampling another 'good' tree.

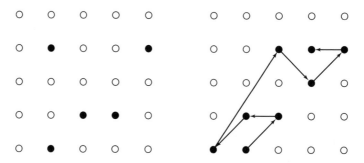

Figure 6.10   Tree space is shown as a two-dimensional grid. In the left figure histories are sampled independent of each other (those samples are black), as in Griffiths and Tavaré (1994). In the right figure, histories are sampled according to a Markov Chain, such that the choice of a new history is dependent on the choice of the previous history.

Recommended reading

Hudson, R. R. (2001*b*), 'Two-locus sampling distributions and their application', *Genetics* **159**(4), 1805–1817.

Stephens, M. and Donnelly, P. (2000), 'Inference in molecular population genetics', *J. R. Stat. Soc. B* **62**, 605–635.

Tajima, F. (1983), 'Evolutionary relationship of DNA sequences in finite populations', *Genetics* **105**(2), 437–460.

Tavaré, S. (in press), *Ancestral Inference in Population Genetics*, Springer Verlag.

Watterson, G. A. (1975), 'On the number of segregating sites in genetical models without recombination', *Theor. Popul. Biol.* **7**(2), 256–276.

7 LD mapping and the coalescent

This chapter aims to introduce concepts that are important for understanding the challenge of finding genes underlying complex diseases through linkage disequilibrium (LD) mapping. It has a bias towards the current use of coalescent theory in the field. Consequently, important research areas such as disease etiology, linkage analysis, and family-based association mapping are only superficially treated.

7.1 The potential of LD mapping

LD mapping is a rapidly growing research field fuelled by the hope that it may lead to the identification of genes and gene variants underlying complex, multigenic traits, such as common diseases (e.g. cancers, Alzheimer's, diabetes, schizophrenia). Classical quantitative genetics and linkage mapping, while being successful in localising genes for Mendelian diseases, have only very limited success in identifying susceptibility genes which are characterised by smaller and context-dependent (genetic and environmental) effects. There is no guarantee that LD mapping will be more successful, but enthusiasm for LD mapping has recently increased because of the large amount of genetic markers that have become available for humans, and soon will be available for a number of other species. More than 5 million SNP markers are now publicly available (see Table 1.1). These can be used for genome-wide scans of association between markers and disease status at densities averaging one marker per kilobase. However, typing is still expensive (even though the price per genotype decreases rapidly), and the common strategy for investigation of a given trait is to localise interesting regions of the genome by more traditional linkage mapping. Pedigree analysis only allows localisation of disease genes to within a 1–2 cM (1–10 Mb) region, dependent on the size (number or observed meioses) of the pedigree. Subsequently, one zooms in on putative causative site(s) using association or LD mapping. This strategy is being employed in several homogeneous and isolated populations, such as the Icelandic, Finnish, and French Canadian populations. As an example, in the Icelandic population the company Decode utilises the quite accurate multigenerational pedigree to identify

Table 7.1 Partial list of genes claimed to be involved in susceptibility to schizophrenia with their chromosomal location (chromosome, and band position)[a]

Gene symbol	Gene name	Chromosomal position
APOL1	Apolipoprotein L, 1	22q12.3
APOL2	Apolipoprotein L, 2	22q12.3
APOL4	Apolipoprotein L, 4	22q12.3
CHRNA7	Cholinergic receptor, nicotinic	15q14
7COMT	Catechol-O-methyltransferase	22q11.2
DAO	D-amino-acid oxidase	12q24
DISC1	Disrupted in schizophrenia 1	1q42.1
DISC2	Disrupted in schizophrenia 2	1q42.1
DRD3	Dopamine receptor D3	3q13.3
DTNBP1	Dystrobrevin binding protein 1	6p22.3
NRG1	Neuregulin 1	8p22-p11
PRODH2	Proline dehydrogenase (oxidase) 2	22q11.2
PRODH	Proline dehydrogenase (oxidase) 1	22q11.2

[a] List is extracted from GeneCards database, August 2003.

interesting regions of the genome for a given trait. This is done using a genome scan with 10,000 microsatellite markers. However, to further zoom in on single genes within such regions, Decode uses a much denser set of SNP markers and LD mapping tools such as the ones discussed below. That such an approach can work was demonstrated recently by the identification of a gene apparently involved in hereditary schizophrenia, the neuregelin gene, see Table 7.1 and Figure 7.20.

7.2   Linkage versus LD mapping

LD mapping should not be confused with linkage mapping, except that the former conceptually can be thought of as population based linkage mapping. Linkage mapping is based on cosegregation of a disease phenotype with a marker genotype in pedigrees. Figure 7.1 shows how a marker allele m_1 cosegregates with disease status. Linkage mapping has successfully been applied for the identification of a large number of rare, single gene human diseases (e.g. see Strachan and Read 2003). Linkage mapping is often efficient in finding markers, which are physically linked to the trait of interest. However, the resolution of the method is determined by the number of meiosis observed in the pedigree, which rarely can be traced with confidence more than a few generations back. (In addition only genetic data from individuals alive today can be obtained.) This implies that a large genomic region around the unknown location of the trait has not experienced any

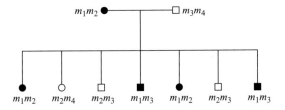

Figure 7.1   Linkage mapping. Search for cosegregation of a marker allele with disease status. In this case a marker locus *M* is typed in an affected female (filled circle) and an unaffected male (empty square) and their children. The allele m_1 is seen to cosegregate with the dominant disease in this example.

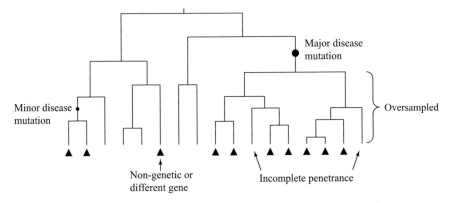

Figure 7.2   An example genealogy for a case-control data set of a complex disease. The mutation increasing risk of the disease may not be unique (i.e. has happend more than once in the sample's history), or several mutations in positions in tight linkage with each other may have occurred. Further the disease may not be fully penetrant. Cases have been oversampled compared to the frequency of the disease in the population. (Figure adapted from McVean, personal communication.)

recombination events in the pedigree and it is thus not possible to determine the location of the trait mutation any more accurately than to somewhere in the region. The region is likely to cover several Mb of DNA sequence and to contain several hundred genes.

LD mapping does not require pedigree information but rather a population sample from affected and unaffected individuals with a number of genetic markers typed at known positions. The rationale is that the unknown genealogy of the sample is a very large pedigree spanning many generations (see Figure 7.2). The basic coalescent process predicts that the genealogy of the sample will be close to $4N$ generations deep, which for human populations might be 4,000–40,000 generations or 100,000–1 million years assuming a generation time of 25 years. Hence it contains far more meioses than the (known) pedigrees underlying linkage mapping, and it becomes possible, in principle at least, to map a trait of interest much more accurately. The problem is of course that the genealogy is unknown and must be modelled statistically. As we have seen the coalescent process depends on various assumptions and parameters which are unknown, and this complicates the analysis of the data further.

Application of coalescent theory in LD mapping increases rapidly (Nordborg and Tavaré 2002; Rosenberg and Nordborg 2002). Obviously, the coalescent process should include recombination/gene conversion, since these forces decouple a marker from a causative mutation. Additionally, for applications to the human population, demographic effects such as population size changes and population structure should be included. One possibility are models in the spirit of the model of Wilson and Balding shown in Figure 4.16. Taking account of oversampling of cases is generally difficult, the exception being the case of a single disease mutation where the coalescent process is well understood. In other cases, heurestic or approximative methods are required.

7.3   Complex disease aetiology

The complexity in locating susceptibility genes for a disease is illustrated in Figure 7.3. The aim is to localise mutations (D_i) that individually or combined affect susceptibility, through a set of markers (M_i). The set of markers may or may not include some of the D_is. The presence of other genes in other areas of the genome also affecting disease status, and properties of the population under study such as admixture, increases the complexity of the task. Finally, environmental factors are often more important for disease status than is genetic variation. Thus, while there are possible great benefits to human health by deciphering genetic susceptibility factors, exercise, diet, and exposure to environmental hazards will remain important determinants of human health.

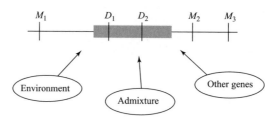

Figure 7.3   Some factors complicating identification of a disease-related gene: Markers M_1, M_2, and M_3 are used indirectly to infer the disease-related gene, which may contain different mutations conferring the disease phenotype (here D_1 and D_2). A number of other factors complicate localisation, including the effect of other genes elsewhere in the genome, population stratification, and the environment. Some of these complicating factors can be directly modelled whereas others are best treated as noise.

In the discussion above, an apparently simplyfying assumption is made: The phenotype (here disease status) of a disease is defined as a binary character that takes the value 1 if an individual has the disease and 0 if the individual is not affected by the disease. Many human disorders are very heterogenous and the phenotype can only superficially be classified as binary. Schizophrenia is an example of a highly heterogenous disorder, as are many other psychiatric disorders, as well as cancers and cardiovascular diseases. Classifying disease status as binary might complicate the search for the underlying genetic make-up unnecessarily; more differentiated phenotypic information could be useful in disentangling different genetic causes of the disease. Thus, dictomising disease status might reduce the chance of locating important genes and gene variants.

In some cases, the heterogeneity of the disease may be revealed by direct means. Many cancers and psychiatric disorders may be subclassified into phenotypically different classes by microarray analysis of mRNA of the affected tissues. In many cases, good classifiers have been built on mRNA data. Furthermore, intrinsic (e.g. sex, age-of-onset) and environmental (e.g. smoking, occupancy, social status) covariates may also provide information on heterogeneity.

One way of representing the relationship between the genetic make-up of an individual and a corresponding phenotype is shown in Figure 7.4. Different combinations of genetic determinants give rise to different phenotypes at the level of the cell. These are classified together as one phenotype at the level of the organism, even though they might be slightly different from each other. One model of a phenotype could be

$$\phi = \phi(g_1, g_2, \ldots, g_k, e_1, e_2, \ldots, e_l), \tag{7.1}$$

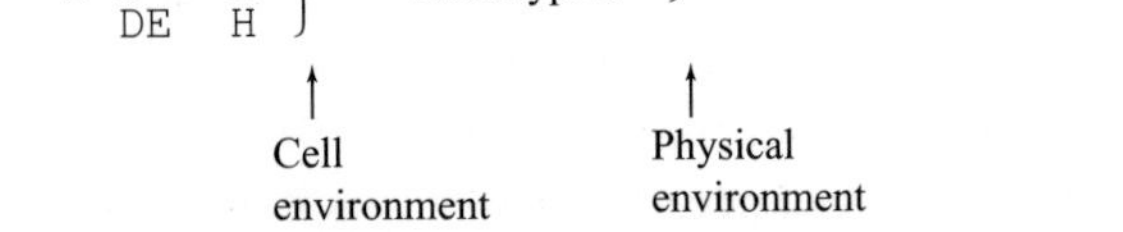

Figure 7.4   Illustration of how sets of different combinations of genetic determinants (here called A, B, ..., H) may lead to the same phenotype in the cell (e.g. affecting the same biochemical pathway) and how these different cell phenotypes may result in the same phenotype for the individual (e.g. the same disease diagnosis).

where $g_1, \ldots, g_k$ denotes the allelic states of k markers affecting the phenotype, and $e_1, \ldots, e_l$ is a series of l environmental factors. For each g_i it is specified which allele comes from which of the two chromosomes. The function ϕ might be taken to be stochastic to allow for influence from other factors and genes that are not included in the model or that cannot easily be modelled, or to allow for penetrance. Thus two individuals who are identical at all g_i and e_js might have different phenotypes due to stochastic effects. As an example, age is an important environmental factor that often influences the onset of a disease, however, it does not determine the onset in a strict deterministic sense. Or a continuous trait such as height is determined by a very large number of factors, including diet and medical history of the person in question, factors that cannot be built into a model easily, unless as stochastic elements.

There does not seem to be general algebraic forms of ϕ that can be used to model an arbitrary phenotype. Often it is assumed that genes have additive effects such that the combined effect of two genes is the sum of the effects of single genes. This is widely believed to be too naive. If we assume two genes (or SNPs) with two alleles each, g_1 and g_2, then there are ten different combinations of genotypes. Assuming no environmental effects and a deterministic binary phenotype this yields a total of $2^{10} - 2 = 1022$ possible mappings of genotypes into phenotypes (discounting 2 because the cases where all ten combinations are mapped to the same phenotype is disregarded). Many of these represent models with epistatic effects which are much more complicated than those models considered in quantitative genetics. Currently it is unknown how complicated the relationship between phenotype and genotype can be.

A single gene (g_1) gives rise to $2^3 - 2 = 6$ different mappings, again assuming no environmental effects and a binary phenotype. Some of these are well-known and well-studied in quantitative genetics, for example, if $\phi(0/1) = \phi(0/0) = 1$ and $\phi(1/1) = 0$ then the 0-phenotype is called dominant; if $\phi(0/0) = 1$ and $\phi(0/1) = \phi(1/1) = 0$ then the 0-phenotype is recessive. (Here x/y indicates that x is on one chromosome, y on the other). The number of mappings increases dramatically if interactions between genetic and environmental factors are allowed.

Many Mendelian disorders, such as cystic fibrosis, can be caused by a series of different mutations in the same gene. The different mutations all have the same effect, namely altering the expression of the same gene. A similar scenario is seen in some complex diseases, for example, there are several hundred mutations in the *BRCA1* and *BRCA2* genes that increase risk of contracting breast cancer. Whether this kind of scenario is prevalent for complex diseases in general is still a matter of investigation. If it is, it will be of huge importance for locating disease related genes: A gene harbouring many potential sites for disease mutations is less easily located than a gene with few potential sites. Individuals affected by the disease become less related compared to a situation where they all share the same mutation. In consequence the individuals are also less likely to be identical genetically in a region around the disease-causing variant.

These complexities discussed above have spawned debate on whether susceptibility genes can be found by LD mapping or any other currently available method. This debate is centred around whether any weak 'signal' of a gene affecting the probability of disease can be found in a sea of 'noise' caused by the environment and other genes (Weiss and Terwilliger 2000). It may be fair to say that there is a general consensus that some genetic determinants can be found (and some have indeed recently been identified with great confidence) whereas others probably cannot be found. The remaining argument is on the relative proportion of each.

Many candidate susceptibility genes for complex disorders are being claimed. A simple search using 'Schizophrenia' as search word in Gene-Cards resulted in (among others) the genes listed in Table 7.1, including neuregulin discussed above. Thus, ample opportunities exist for adopting candidate gene approaches to LD mapping rather than genome wide scans. However, in either case efficient analysis will be complex and may benefit from methods such as the ones discussed below.

Much more empirical research and methodological development is needed before it is clear whether LD mapping can live up to optimistic expectations, but it is clear that optimal use of the data and considerations of all important forces (demographic, selection, model of recombination, etc.) is demanded to have any chance of distinguishing a real signal from noise. Explicit modelling of disease models allowing for epistasis among unlinked genes is only in initial stages of development, and the amount of power gained by doing such modelling is unclear.

7.4 Formulating the task

With the definition of a phenotype model in the previous section it is possible to formulate the task of LD mapping in a formal way. Define the possible relationship between phenotype, genotypes, and environmental effects through equation (7.1). To allow for some flexibility and the fact that we do not want to fix potential interactions between the putative phenotype related genes in advance, we might define a class of phenotype functions. For example, if we postulate two phenotype related genes (or markers) and one environmental factor, the phenotype could be defined as

$$\phi(G + e) = 1\{G + e > 0\}, \qquad (7.2)$$

where e is the level of the environmental factor, $1\{x > 0\}$ is an indicator variable that takes the values 1 if $G + e > 0$ and 0 otherwise, and G is

$$G = \alpha_1 \#\{0 \text{ alleles in site 1}\} + \alpha_2 \#\{0 \text{ alleles in site 2}\}. \qquad (7.3)$$

Here α_i are undefined numbers. This definition of ϕ puts every individual in either one of two classes, $\phi = 0$ or $\phi = 1$. The effect of the two genes is a weighed average of the number of 0 alleles.

In addition to observing e a set of markers is also observed. These may or may not include the two putative phenotype related genes. Our task is to find the best (in some sense) location of the two genes and the values of α_1 and α_2. In this example no epistatic effects are included and if, for example, α_2 turns out to be zero, only the first gene has an effect.

The example that has received most attention in the literature, and that undoubtly is the most simple example possible, is given by

$$\phi(0/0) = 1, \quad \text{and} \quad \phi(0/1) = \phi(1/1) = 0. \qquad (7.4)$$

That is, only one putative gene, no environmental effects, and all homozygous for the 0 allele are affected and no one else. If the frequency of the 0 allele is small in the population, and affected are overrepresented in a sample relatively to the population frequency, it is likely that only homozygous, either 0/0 or 1/1, are present in the sample, because heterozygous become very rare. This has the advantage that it is known whether a haplotype carries the 0- or 1-allele.

7.5   A role for the coalescent

Coalescent theory has at least two major roles to play in LD mapping. These are:

1. Generation of simulated data. Adopting a model of the phenotype such as equation (7.1) and a model of the coalescent process, samples of random and phenotype related (the g_is in equation (7.1)) markers can be simulated. Such samples can be used for validation of methods for localising genes and gene variants; in particular it is possible to investigate the effects of marker density, marker frequencies, sample sizes, epistasis, and so on. Irrespective of the true genotype to phenotype function, the foundation for any gene mapping method is correlation of genealogical histories of markers at close positions. Simulations can also (as shown in the next section) be used to visualise how recombination and gene conversion influence these.

Simulation progresses in several steps. First, the genealogy conditional on the location (and type) of the g_is is simulated. This can be done in a related way to equation (3.43) and equation (6.19) using Bayes' formula. Second, random mutations are placed on the genealogy. Third, the phenotype is determined from the g_is and potential environmental factors (that also might be simulated according to a stochastic model). Individuals are made by randomly pairing sequences. Lastly, a subset of the random markers and/or individuals can be selected to mimic experimental design criteria such as: all markers have minor allele frequency above some threshold, exactly m markers are typed, the ratio of cases to controls is fixed at some value, and so on.

2. Analysis of real data: Using coalescent theory explicitly to analyse data. Such methods will be discussed briefly at the end of the chapter. Here we formulate the problem more generally. Again we need a model of the phenotype. Assume for a start that the positions of the phenotype related markers, g_i, are known relative to the other markers, m_i. Also assume that the phase of all markers are known, that is, we have $n = 2(n_1 + n_2)$ haplotypes available, where n_1 is the number of cases and n_2 the number of controls. For any given combination, G, of alleles at the phenotype related markers we can evaluate the probability of obtaining G given the observed markers m_i and phenotypes under a given coalescent model. This is done by calculating the probability of obtaining G, m_i, and the phenotypes and dividing this quantity by the probability of m_i and the phenotypes (this is just the definition of a conditional probability). The latter probability involves summing over all possible combinations of G;

$$P(G \mid m_{ij}, \phi_j, j = 1, \ldots, n, i = 1, \ldots, m)$$
$$= \frac{P(G, m_{ij}, \phi_j, j = 1, \ldots, n, i = 1, \ldots, m)}{P(m_{ij}, \phi_j, j = 1, \ldots, n, i = 1, \ldots, m)}, \tag{7.5}$$

where

$$P(m_{ij}, \phi_j, j = 1, \ldots, n, i = 1, \ldots, m)$$
$$= \sum_G P(G, m_{ij}, \phi_j, j = 1, \ldots, n, i = 1, \ldots, m). \tag{7.6}$$

These probabilities will in general be intractable but can be calculated using MCMC and IS methods, as discussed previously. The combination G and the positions of g_i that provide the highest probability might be selected as representing the optimal choice given the model and the observations at hand. Environmental effects can in principle be included, adding them alongside the ϕ_js. Also multistate and continuous phenotypes can be dealt with in this way, although the computational burden might increase dramatically. If the phenotype is continuous the probabilities in the bottom equation become replaced by densities.

Below we will show how simulation can be used to understand aspects of the genealogy surrounding a putative disease mutation. It is straightforward to simulate recombination, gene conversion, exponential growth, and population structure in a single coalescent process, treating waiting times as the outcomes of competing, independent, exponentially distributed variables, extending, for example, Algorithm 3 of Chapter 2 accordingly.

7.6 Genealogical trees around a disease mutation

Focus on the simple situation where a single variant causes an increased risk of disease. We term the position of the variant the target. A well-studied example of a single target mutation is the Apolipoprotein E gene discussed in Chapter 5. Association of a set of SNP markers around the target (a SNP itself) was investigated by Martin et al. (2000) and their results are shown in Figure 7.5. The most significant association is found, not surprisingly, at the causative site, but a number of linked sites also show association suggesting that LD mapping may be feasible for markers within this spacing. However, more surprisingly, some of the very close markers show very little association. Whether this is caused by the highly stochastic nature of the combined action of the recombination process and

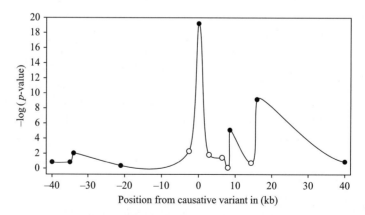

Figure 7.5 Association of thirteen SNP markers with susceptibility to early onset of Alzheimer's disease. The causative SNP shows the strongest association, but other close markers also show strong association due to LD with the causative marker. Several markers in between (marked by open circles) show virtually no association. (Data from Martin et al. 2000.)

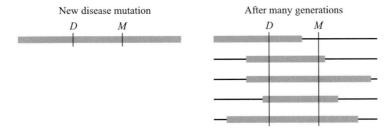

Figure 7.6   LD. A new mutation, D, contributing to a disease initially enters a population through mutation on a specific genetic background. After many generations recombination has decoupled D from its initial genetic background except for a (small) region around D. A marker, M, in this region is in LD with D.

the mutation process creating the markers, gene conversion unlinking these markers from the other markers, or bias in the LD mapping method is not known. However, this example motivated us to include gene conversion in the following investigation.

The statistical power of LD mapping depends on the number of markers near the target that also are associated with the disease, though not causative. For a marker in a given region to be informative about the target site, the marker needs to share some evolutionary history with the target. A given mutation contributing to disease risk will arise on a unique genetic background, and initially be completely linked with any other marker M in this background. Recombination breaks linkage down over time. Theoretically, LD is expected to decay by a fraction $1 - r$ every generation, where r is the recombinational distance between D and M per generation. Figure 7.6 exemplifies how recombination may break down the initial association between M and D. Gene conversion would have a different effect in removing association for all but a small fragment of the chromosome.

Viewed from a coalescent perspective, association between D and M is caused by correlation between the genealogical trees of the target and the marker site. A strong correlation is expected if the recombination rate is low, and a weak correlation is expected for sites further apart. A correlation between genealogical trees manifests itself through an overall 'similarity' of the trees. How can we measure similarity of coalescent trees? Different qualitative and quantitative measures can be used.

7.6.1   Qualitative measures

Examples of qualitative measures include:

1. Two genealogical trees are identical (equal branch lengths and topology).
2. Two genealogical trees have the same topology.
3. Two genealogical trees share MRCA.

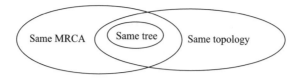

Figure 7.7   Venn diagram of the three qualitative measures of tree similarity.

Note that if a tree belongs to class 1 it also belongs to class 2 and 3, (but not vice versa) and that a tree can belong to class 2 without belonging to class 3 and vice versa (Figure 7.7).

7.6.2   An example

The coalescent with recombination and gene conversion can simulate an ancestral recombination graph for a sample of sequences of length $\rho/2$. We may focus on a specific point on the sequences, the presumed target point. In this target point, a specific coalescent tree exists. A (stochastic) neighbourhood around this target will not have experienced any recombination events in the history of the sample and will therefore have the same tree. Moving further away from the target, recombination events will create trees different from, but correlated with, the target tree. Figure 7.8 illustrates three qualitative measures for ten random outcomes of the coalescent process for twenty genes, and a scaled recombination rate of $\rho = 2$. This amount of recombination corresponds to approximately 1–100 kb in the human genome, depending on the local rate of recombination.

The figure shows that the size of the region with the same tree is highly variable, but that it is generally a continuous segment around the target that shares the same tree as shown in Figure 7.6.

The example illustrates well the very large amount of variation in the three measures, reflecting the variance in the coalescent process. The large variance implies that LD mapping might require a much denser map of markers in some regions than in others even when the underlying evolutionary parameters are the same.

Figure 7.9 shows the effect of adding gene conversion to the coalescent process of Figure 7.8. Gene conversions occur at a scaled rate of $\gamma = 8$, with an average tract length $1/Q = 1/25$ of the region. The scaled rate of gene conversion is four times the rate of recombination, chosen to be within the estimated range of 2–10 reported by Frisse et al. (2001).

Figure 7.9 shows that gene conversion fragments the region around the target with the same tree, topology, or MRCA. Figure 7.10 shows the number of disjoint segments with the same tree with and without gene conversion averaged over many runs. Thus, there are regions with trees similar to the target, separated by regions with less similar trees. Gene conversion

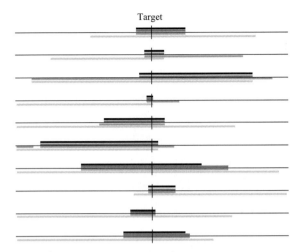

Figure 7.8 Ten examples of the structure of the genealogy related to the target position in a sample of twenty sequences. The tree of the target position is also the tree of positions in a neighbourhood around the target point, where no recombination events have happened (top line, in black). In dark grey is marked regions with the same tree as the target (second line from top); irrespective of recombination events or not. It includes the region of the black line and potentially more because of invisible recombination events. In grey is marked regions where the tree of the sample has the same topology as the target (third line from top). Mutations in this region are all compatible with the same tree. In light grey (bottom line) are the regions that share MRCA with the target position. In each case, results are based on a sample of twenty genes, and a scaled recombination rate of $\rho = 2$.

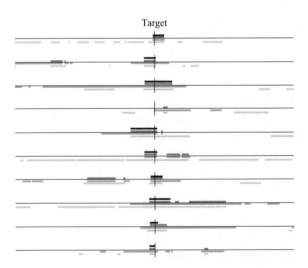

Figure 7.9 The effect of including gene conversion. The scaled recombination rate was set to $\rho = 2$ as in Figure 7.8, the scaled gene conversion rate $\gamma = 8$, and the average gene conversion tract length was set to 1/25 of the length of the sequence.

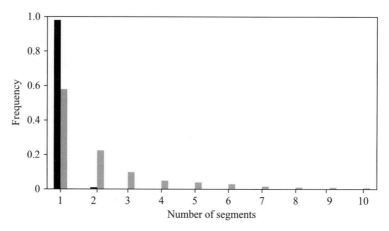

Figure 7.10   The number of segments with the same tree as at the target position for two situations: with (grey columns) and without (black columns) gene conversion.

Table 7.2   The probability of observing a segment larger than the target segment for each of the three qualitative measures, with and without gene conversion[a]

Recombination and gene conversion rates	$\rho = 2, \gamma = 0$	$\rho = 2, \gamma = 8$
No. of segments with same tree	1.02	1.8
P(target segment not largest) (%)	0.2	14
No. of segments with same topology	1.02	2.1
P(target segment not largest) (%)	0.3	20
No. of segments with same MRCA	1.1	2.9
P(target segment not largest) (%)	1.5	25

[a] Results are based on 10,000 replicate simulations.

changes the genealogical history of a small segment of the sequence as opposed to recombination, which leads to different genealogical histories for the left and right part of the sequence (seen from the recombination break point). Clearly, the fragmentation of segments with similarity to the target also has implications for the success of LD mapping, in particular if the fragment containing the true target point is not the largest fragment. In such cases, using LD mapping, one would be more likely to locate the target on one of the fragments not carrying the true target. At the same time, one would risk finding regions at either side of the 'false' fragment with little association to the disease, increasing one's confidence that the fragment identified indeed contains the target. The probability of finding a false fragment which is larger than the target fragment is shown in Table 7.2 for the same parameter set as in Figure 7.9, and compared to the situation with no gene conversion (i.e. Figure 7.8). Clearly, large false fragments do occur at an appreciable frequency.

7.6.3   Quantifying genealogical tree differences

It is also of interest to try to measure how different the genealogical trees become when moving away from the target site. Several measures may be considered.

The first measure is the probability that a mutation in each of the trees would create the same bipartition of the sequences (into ancestral and mutant types). Figure 7.11 shows two correlated, but topologically different trees, whose bipartitions are the same.

The amount of shared bipartitions is directly relevant for the interpretation of SNP data discussed below, because it predicts the probability that mutations on the two different trees create the same sample configuration and thus are in strong LD. This is termed 'perfect' LD .

This leads us directly to our first measure: find the set of branches in two trees that give identical bipartitions of the sequences, and weight these by the relative lengths of the branches, that is,

$$M_{AB} = \frac{\sum_{i,j} I_{\{i=j\}} a_i b_j}{l_A l_B},$$
(7.7)

where a_i (b_j) is the length of branch i (j) in tree A (B), $I_{\{i=j\}}$ is an indicator function which is one if branch i in A and branch j in B give the same bipartition, and l_A (l_B) is the total branch length of tree A (B). It is reasonable to compare M_{AB} with M_{AA}, the similarity of a tree to itself. We have

$$M_{AA} = \frac{\sum_{i,j} a_i^2}{l_A^2},$$
(7.8)

and define the ratio by

$$S_{AB} = \frac{M_{AB}}{M_{AA}}$$
(7.9)

(note that S_{AB} is not symmetric in A and B). The measure is heavily influenced by the terminal branches of the tree, which always give identical

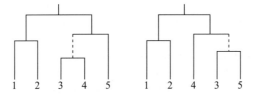

Figure 7.11   Two different trees for five sequences, differing by at least one recombination event. Dashed lines correspond to bipartitions not found in the other tree.

bipartitions in two trees, so equation (7.7) might be refined by considering only bipartitions corresponding to polymorphic sites with minor allele frequency greater than some threshold q, for example, $q > 0.1$. This has the added advantage of facilitating comparison to empirical studies, where it is custom practice to use only markers with a minor allele frequency above a certain threshold, typically $q \approx 10$–20%. M_{AB} is closely related to the r^2 measure of LD discussed below, in that it determines the probability of observing markers with $r^2 = 1$.

Figure 7.12 shows the behaviour of S_{AB} for the same parameters used in Figure 7.9, without and with gene conversion. The full lines show the average over 1,000 runs, whereas the dotted lines shows the results from two randomly chosen runs. Without gene conversion, S_{AB} decreases more regularly with increasing distance from the target than with gene conversion, but in both cases less informative trees can separate a more informative tree

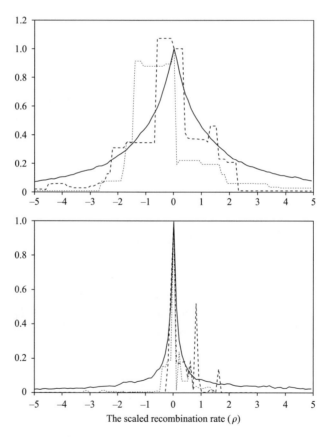

Figure 7.12 S_{AB} over the sequence of the ancestor to the target. The average over 1,000 replicates is shown by the full line, dotted lines are two randomly chosen runs. (Top) No gene conversion, $n = 50$, $\rho = 10$, $q = 0.1$. (Bottom) Gene conversion added to (Top) with rate $\gamma = 40$ and tract length parameter $Q = 25$.

from the target tree. The average S_{AB} decreases rapidly away from the target when gene conversion is present, since gene conversion works primarily over short distances. A somewhat counterintuitive observation with some practical importance is that there often exist trees with higher probability of markers being in perfect LD with the target tree than the probability of markers being in perfect LD at the target tree (i.e. $S_{AB} > 1$). This can occur when relatively long branches in the target tree are even longer in the tree it is compared to.

Another possible measure is the difference between two trees, A and B, as measured by the sum of the differences of branch lengths leading to the same bipartition relative to the total branch lengths. That is,

$$N_{AB} = \sum_i \left(\frac{a_i}{l_A} - \frac{b_i}{l_B} \right) = 1 - \sum_i \frac{b_i}{l_B}, \qquad (7.10)$$

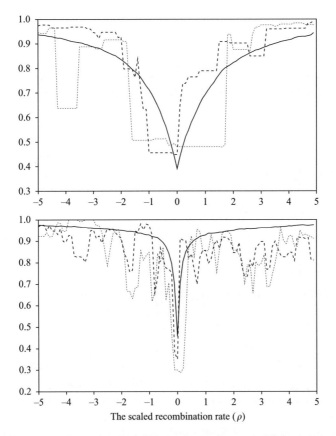

Figure 7.13 N_{AB} over the sequence of the ancestor to the target. The average over 1,000 replicates is shown by the full line, dotted lines are two randomly chosen runs. (Top) No gene conversion, $n = 50$, $\rho = 10$, $q = 0.1$. (Bottom) Gene conversion added to (Top) with rate $\gamma = 40$ and tract length parameter $Q = 25$.

where $b_i = 0$, if no branch in tree B induces the same bipartition as branch i in the target tree A. This measure can also be restricted to a given frequency q of bipartitions. Clearly, if the topology of two trees is the same, they induce the same set of bipartitions and $N_{AB} = 0$.

This measure was calculated with a similar set of parameter values as used in Figure 7.12, and Figure 7.13 shows the results. Apart from being a measure of difference rather than a measure of similarity, the N_{AB} measure shows the same trend as S_{AB}: Regions with similar trees are interrupted by regions with very different trees.

7.7 The genealogical process reflected in data

The previous discussion has aimed at characterising aspects of the genealogical trees over the sequences, as a function of recombinational distance from a given target. These trees are never directly observable. Instead we have to rely on genetic markers to get as much information as possible about the tree in each position. However, since we have a limited number of markers, mapping with markers has a further element of noise inherited from the genealogical process than if we had perfect knowledge about the trees.

SNPs are the preferred marker type for LD mapping because of their abundancy in the genome, see Table 1.1. However, large differences in the density of SNPs exist over the genome (Venter et al. 2001; Reich et al. 2002). This difference can be due to:

1. Variance in the genealogical process.
2. Differences in the per base pair recombination and gene conversion rates over the genome.
3. Differences in mutation rate over the genome.
4. Differences in the selective regime over the genome.

Most is presently known about the magnitude of the first effect, which is relatively straightforward to characterise using simulation under a demographic model. Recent studies utilised the divergence between human and chimpanzee on human variation data to suggest that differences in recombination rates are more important than differences in mutation rates (Reich et al. 2002).

A convenient feature of SNP markers is that they are very likely to be caused by a single mutation. This is because the per site mutation rate is on the order of $u = 5 \cdot 10^{-9}$ per generation. Assuming a human effective population size of 10,000, this implies that the scaled mutation rate for a given site is $\theta = 4Nu = 2 \cdot 10^{-4}$ per base pair. In a sample of fifty humans we would thus expect $\theta \sum_{i=1}^{49} 1/i \approx 10^{-3}$ mutations (equation 2.32), and it is unlikely that we observe two mutations in the same site. Thus, assuming a SNP represents a single mutation, it creates a unique observable bipartition of the sequences into those having the first allele and those having the second.

Normally, it is not possible to tell which of the two variants at a given SNP is the ancestral type. However, the most frequent variant is also most likely the oldest. Under the basic coalescent, the probability that a given variant is the oldest is equal to its frequency. If we do not know which allele is ancestral, then we may instead speak of the age of the polymorphism as a function of the frequency of one variant.

However, if the homologous region is sequenced in chimpanzee (or another close relative to man), the ancestral type can often be determined. It will then be possible to estimate the age of the derived allele (the mutation) from equation (3.38). Knowledge about which variant is derived is valuable information in LD mapping, because LD (see below) is expected to extend further around the derived allele if the LD is primarily caused by a new mutation combined with genetic drift. This is one of the considerations one may have when choosing which SNP markers one should use from a larger set.

7.8 Linkage disequilibrium (LD)

One measure of LD is

$$D = p_{11}p_{22} - p_{12}p_{21} = p_{11} - p_1q_1, \tag{7.11}$$

where p_{ij} is the proportion of gametes carrying allele i at site 1 and allele j at site 2 and p_i and q_i are the frequencies of allele i at site 1 and 2, respectively. If the two sites evolve independently of each other, for example, if they are far apart, then $p_{ij} \approx p_iq_j$ is expected and consequently $D \approx 0$.

A large number of alternative measures of LD between two markers has been proposed. Commonly applied measures take D and normalise it in some way, for example,

$$D' = \begin{cases} \frac{D}{\min(p_1q_2,p_2q_1)} & \text{if } D > 0, \\ \frac{D}{-\min(p_1q_1,p_2q_2)} & \text{if } D < 0 \end{cases} \tag{7.12}$$

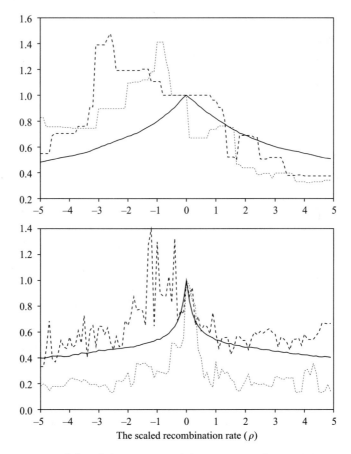

Figure 7.14 Decay of r^2 with distance around the target site. The average over 1,000 replicates is shown by the full line, dotted lines are two randomly chosen runs. (Top) No gene conversion, $n = 50$, $\rho = 10$, $q = 0.1$. (Bottom) Gene conversion added to (Top) with rate $\gamma = 40$ and tract length parameter $Q = 25$.

and

$$r^2 = \frac{D^2}{p_1 p_2 q_1 q_2}. \tag{7.13}$$

Continuing our example, Figures 7.14 and 7.15 show the average LD with the target tree over a sequence for both D' and r^2 as a function of recombinational distance, with and without gene conversion. D' and r^2 are calculated from the genealogical tree in each position, by calculating LD for each combination of branches, and weighing the sum by the relative length of each branch (using a threshold frequency of $q = 0.1$).

The r^2 and D' measures differ in their properties. Figure 7.16 shows three sample configurations for two sites. If two markers are in equal frequency

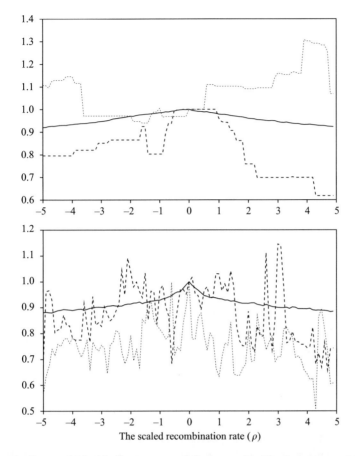

Figure 7.15   Decay of D' with distance around the target site. The average over 1,000 replicates is shown by the full line, dotted lines are two randomly chosen runs. (Top) No gene conversion, $n = 50$, $\rho = 10$, $q = 0.1$. (Bottom) Gene conversion added to (Top) with rate $\gamma = 40$ and tract length parameter $Q = 25$.

1	A	C	$r^2 = 1$	A	C	$r^2 < 1$	A	C	$r^2 < 1$
2	A	C	$D' = 1$	A	G	$D' = 1$	T	C	$D' < 1$
3	T	G		T	G		A	G	
4	T	G		T	G		T	G	

Figure 7.16   The behaviour of r^2 and D' for the three cases with two, three, and four haplotypes in a sample of four chromosomes. $D' < 1$ only when the number of haplotypes is four.

and create only two haplotypes, both measures attain their maximum value of 1. This situation would be created if the genealogical trees in the two positions were topologically identical and mutations occurred in corresponding branches. If markers are in unequal frequency, at least three haplotypes are present. If exactly three haplotypes exist, $D' = 1$, whereas $r^2 < 1$. This situation could be created if the genealogical trees at the two positions were identical, but mutations occurred in different branches. Finally, Figure 7.16 shows the situation where all four haplotypes are present and D' and r^2 are both <1.

7.8.1 Testing for LD

While a measure of LD is informative, we are often more interested in knowing whether two sites are significantly associated, that is, whether the LD detected is statistically significant. The null hypothesis to be tested is whether sites are independent, that is, $H_0 : p_{ij} = p_i q_j$. This can be done in a fairly simple manner from the counts of each set of alleles occurring together. There are four combinations of alleles at the two positions, and the null hypothesis states that the two loci are independent. The natural test is therefore a test for independence of a 2×2 contingency table with elements n_{ij}, where n_{ij} is the number of chromosomes carrying marker i at position 1 and marker j at position 2, and $\sum_{i=1}^{2} \sum_{j=1}^{2} n_{ij} = n$ is the total sample size.

The most precise test of this appears to be Fisher's exact test of independence. However, Fisher's exact test is complicated to calculate when n is large, and we may instead calculate

$$X = \sum_{i,j} \frac{(n_{ij} - e_{ij})^2}{e_{ij}}, \tag{7.14}$$

where e_{ij} is the estimated expected number of n_{ij} under the assumption of no association, $e_{ij} = (n_{i1} + n_{i2})(n_{1j} + n_{2j})/n$. If all e_{ij} are sufficiently large (say greater than five), then X is approximately χ^2 distributed with one degree of freedom. Figure 7.17 shows how the relative probability of observing significant LD decays with distance.

7.8.2 Accounting for population admixture

Some problems with population structure can be alleviated by making sure that the sample is from a single population only. However, many human populations (e.g. African-Americans) are true mixtures of populations. Methods exist that use unlinked markers to control for this situation, either by explicitly modelling admixture in the sample (the population assignment approach by Pritchard et al. 2000), or by controlling for it (the genomic control approach by Devlin and Roeder 1999). Population assignment uses unlinked markers to estimate which proportion of genetic material in a given individual has originated from each of a number of populations. The genomic control method adjusts the significance levels of association by modifying the χ^2 distribution by multiplying with a factor estimated from data.

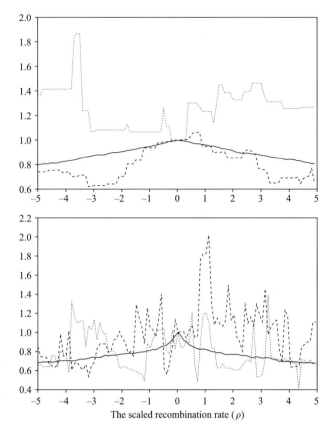

Figure 7.17   The relative probability (compared to the target tree) of being in significant LD between a random marker at the target and anywhere over the sequence, as determined by the χ^2 test. (Top) No gene conversion, $n = 50$, $\rho = 10$, $q = 0.1$. (Bottom) Gene conversion added to (Top) with rate $\gamma = 40$ and length parameter $Q = 25$.

7.8.3   Differences between human populations

Great effort is currently being put into sequencing (or SNP typing) genes in samples of individuals from different human populations in order to search for differences and to choose the most appropriate population for disease mapping. As an example, Figure 7.18 is based on the SeattleSNP survey of almost 16,000 SNPs in 150 different genes which have been typed for SNPs in twenty-four African-Americans and twenty-three individuals of European descent. It is clear that more diversity is generally found in populations of African descent in accordance with the out-of-Africa hypothesis of human descent, see Chapter 8. Furthermore, Figure 7.19 shows that different evolutionary forces may have shaped the variation in these two populations, the European population has a higher average value of Tajima's D and a greater variance. The greater variance may suggest a stronger deviation from demographic equilibrium (e.g. occurrence of a bottleneck) in the population of European descent. This is consistent with

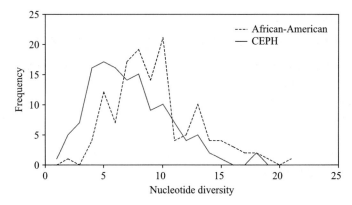

Figure 7.18 The distribution of nucleotide diversity in 150 genes in samples from an African-American population and from the CEPH reference panel. Data are from the Seattle SNP project surveying variation in 150 candidate genes for inflammatory responses in humans (SeattleSNPs. NHLBI Program for Genomic Applications, UW-FHCRC, Seattle, WA (URL: http://pga.gs.washington.edu) accessed January 2004).

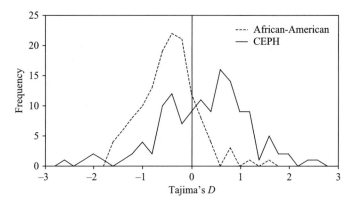

Figure 7.19 The distribution of Tajima's *D* values for the same data set as in Figure 7.18.

the general finding that LD extends for longer distances in European samples compared to African samples. Thus, the amount of recombination in the history of a typical gene in Europeans is smaller than the amount of recombination in an African population. This is likely due to a difference in effective population size, rate of population expansion, population structure, and perhaps selection. It suggests that African populations are most appropriate for very fine-scale mapping (more variation and more recombination events), but that LD mapping is feasible using markers further apart when using a European population. Small isolated populations as found in Sicily, Iceland, Canada, and certain areas of Finland have even more LD than the European population and have therefore been the populations of choice for identifying candidate regions when using a sparse set of markers.

7.9   Measuring association using single markers

Association mapping aims at searching for any association between markers and the phenotype of interest. It does not require any knowledge about dominance, recessivity or penetrance of susceptibility genes in causing the phenotype, but the power of the method is of course dependent on these factors.

The basis for a test of association of a marker allele m at the marker locus M for a binary disease trait is the null hypothesis

$$P(m \mid \text{disease}) = P(m \mid \text{non-disease}). \qquad (7.15)$$

If we think of disease status as a 'marker' then this test is equivalent to the test introduced in equation (7.14). If this hypothesis is rejected, m is said to be associated with the disease. The assumption underlying LD mapping is that the main cause of such association is LD and not one of the other factors discussed above. It is important to remember that association does not imply causation. Strachan and Read (2003) exemplifies this by the strong correlation in San Fransisco between a certain MHC variant and the ability to eat with chopsticks. Clearly, this association is caused by a large difference in frequency of this marker between the Asian-American population and the remaining population of San Fransisco and not because of LD between the MHC variant and a 'chop stick' gene.

7.10   Haplotype LD mapping

Figure 7.20 shows three haplotypes based on microsatellites and SNPs which were found to be significantly overrepresented in schizophrenia patients in Iceland and Scotland (Stefansson et al. 2003). None of the markers are individually strongly associated with schizophrenia. The result suggests that a whole chromosomal fragment around a disease mutation has been identified and obviously statistical power has been gained by considering whole haplotypes consisting of several adjacent markers rather than single markers one at a time.

Haplotype association is generally not as simple as depicted in Figure 7.20. The example considered in Figure 7.5 found strong associations for three markers separated by uninformative markers. Haplotypes consisting of close, non-adjacent markers may therefore also be considered.

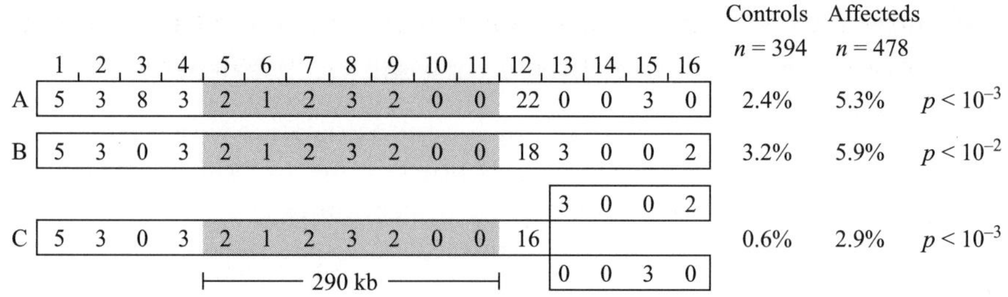

																	Controls	Affecteds	
	1	2	3	4	5	6	7	8	9	10	11	12	13	14	15	16	n = 394	n = 478	
A	5	3	8	3	2	1	2	3	2	0	0	22	0	0	3	0	2.4%	5.3%	$p < 10^{-3}$
B	5	3	0	3	2	1	2	3	2	0	0	18	3	0	0	2	3.2%	5.9%	$p < 10^{-2}$
												3	0	0	2				
C	5	3	0	3	2	1	2	3	2	0	0	16					0.6%	2.9%	$p < 10^{-3}$
			⊢	—	290 kb	—	⊣			0	0	3	0						

Figure 7.20 Haplotypes associated with schizophrenia at the neuregulin gene. Three multilocus haplotypes are highly significantly overrepresented in affected persons (χ^2 test of association). These three haplotypes share a core haplotype (shaded) of seven markers covering a region of 290 kb within which the causative mutation(s) is expected to reside (data from Stefansson et al. 2003).

7.11 Model based LD mapping

Common for the tests discussed above is that they assume an underlying fixed (but unknown) population frequency of marker alleles, haplotypes, and disease mutation. The similar tests cast in a coalescent framework would assume that the frequency, for example, of a marker allele, is a random variable whose distribution is given by an appropriate coalescent model. One would then ask for the posterior density of frequencies given the observed data and enquire whether this density is similar to the density obtained assuming sites are independent or unlinked to a disease causing variant. This approach has much in common to many multipoint LD mapping methods currently available.

Model-based LD mapping as implemented in likelihood or Bayesian multipoint LD methods represents an alternative to the haplotype association approach discussed above. Recent approaches make use of coalescent theory. The basic idea is that a model of the recombination process and of the genealogy of the case chromosomes is assumed. Given genetic data for a set of case haplotypes and a set of controls the methods output a probability density for the location of a disease mutation. This is done in a Bayesian way by assuming prior distributions of parameters (for disease location this could be a uniform prior over the region covered by markers). The posterior distributions can then be found by Bayes' formula:

$$f(\text{parameters} \mid \text{data}) = \frac{P(\text{data} \mid \text{parameters}) f(\text{parameters})}{P(\text{data})}, \qquad (7.16)$$

where $P(\text{data} \mid \text{parameters})$ is the likelihood of the data given a set of parameters.

If one is interested in the distribution of just a subset of the parameters, called x (e.g. the disease position), integration is then done over the remaining parameters

$$f(x \mid \text{data})$$

$$= \int_{\substack{\text{parameters} \\ \text{except } x}} f(\text{parameters} \mid \text{data}) \, d(\text{parameters except } x). \quad (7.17)$$

This integration is usually performed using MCMC sampling (see Chapter 6). Although this is in principle straightforward, the space to sample is enormous, rendering this approach computationally very demanding.

Below is a short description of four recent models with most emphasis on the method by Morris et al. (2002). The methods mainly differ in the genealogical model relating case chromosomes and ability to handle disease heterogeneity.

7.11.1   Star shaped genealogy

7.11.1.1   *Approach of Liu et al.*

Liu et al. (2001) modelled the genealogy of case chromosomes as a star tree. Their approach allows for some disease heterogeneity. Except for the star shaped genealogy, the method is very similar to the method by Morris et al. (2002) discussed below. It does allow for more than one disease mutation, as long as the number of disease mutations is explicitly chosen. In this case, the genealogies for cases descending from each disease mutation are assumed star-shaped but with different MRCA, estimable from data.

7.11.2   Coalescent based genealogy

7.11.2.1   *Approach of Rannala and Reeve*

In the model of Rannala and Reeve (2001), the case chromosomes are considered to be related by a coalescent tree in the unknown position x of the disease mutation. This coalescent process may include demographics such as population expansion and admixture.

The method assumes that markers in the background (control) population are independent of each other, that is, the frequency of a haplotype in the background population is given by the product of allele frequencies over single markers. However, LD in the background population may be modelled; for example, pairwise LD can be modelled by a first-order Markov chain along the chromosome. Further, the method assumes that case chromosomes have all descended from the same disease mutation.

7.11.2.2 *Approach of Morris et al.*

Morris et al. (2002) aimed at combining a realistic model of case chromosome genealogy with the ability to treat multiple disease mutations. They did so by means of a 'shattered' coalescent genealogy as illustrated in Figure 7.21. The shattered coalescent is a coalescent where random branches have been deleted, thus disconnecting the sample. Each of the disjoint trees can then be thought of as representing the trees of different disease mutations. The amount of shattering is determined by a shattering parameter ρ (their notation, different from our scaled recombination rate) that is given a prior distribution. Thus, even though the coalescent process is utilised, the shattered coalescent tree is not a proper coalescent description of the genealogical tree of a set of cases when more than one disease mutation exists. The proper coalescent process for this situation has not yet been derived.

Given case and control chromosomes, Morris et al. aimed at integrating

$$f(x, h, \omega, T, z, N, \rho \mid A, U) \sim P(A, U \mid I, x, h, \omega, T, z, N) \qquad (7.18)$$
$$\times f(\omega, T, z \mid \rho) f(\rho)$$

over all parameters except the disease position x. The meaning of the parameters of the model is shown in Table 7.3. Integration over the shattered coalescent tree (parameterised by T, z, ω) is done by assuming specific haplotypes at internal nodes I and then integrating over all possibilities of I, weighing by their posterior probabilities. This is easily done using MCMC techniques. The main practical problem is the very high dimensionality of the parameter space and how a good proposal distribution is defined for updating of parameters in the MCMC chain. Morris et al. suggested to change one parameter at a time by a random amount within

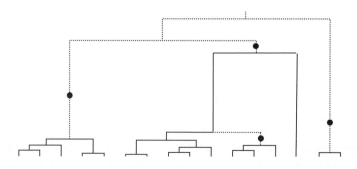

Figure 7.21 The shattered coalescent tree used as the prior on the tree relating case chromosomes in the method of Morris et al. (2002). The tree is built using a coalescent algorithm, but afterwards a number of branches determined by the shattering parameter are removed to model disease mutations with independent origin.

Table 7.3   The key parameters in the approach of Morris et al. to Bayesian multipoint LD mapping

Parameter	Meaning
x	Location of disease locus
h	Population marker-haplotype proportions
ω	Branch lengths of genealogical tree
T	Topology (branching pattern)
z	Parental status
N	Effective population size
ρ	Shattering parameter
A	Cases (affected)
U	Controls (unaffected)
I	Haplotypes at internal nodes in the case tree

certain limits. Changes to the tree can, for example, be moving a node up or down or moving a subtree to another position. As long as the proposals are reversible and irreducible, any proposal function will do and it is up to experimentation to ensure efficient mixing of the Markov chain.

7.11.2.3   *Approach of Larribe et al.*

Larribe et al. (2002) provides a full coalescent approach to inference on the disease position. The method is based on making inference in the ancestral recombination graph of all cases and control chromosomes using recursions similar to Griffiths and Marjoram (1996) as discussed in Chapter 5 and 6. The disease position is then treated as missing data. Importance sampling is used to estimate the likelihood curve for the disease position.

The application of this approach is presently restricted by the assumption of a single disease mutation and by computational intractability, so it is not really amenable to real data analysis. However, there is no reason to believe that further development of methods cannot remove these restrictions.

7.11.3   An example

Figure 7.22 shows an example analysis using multipoint LD mapping on a case-control data set for breast cancer. There are 850 cases and 850 controls measured for eight microsatellite markers. The upper panel shows the association of each single marker from simple χ^2 measures of association. The lower panel shows the posterior distribution of disease located obtained using multipoint LD mapping as implemented in the program GeneRecon which is based on a combination of the above methods. The position of a known candidate gene is shown below the figure and is seen to be in good accordance with the prediction from the Bayesian multipoint

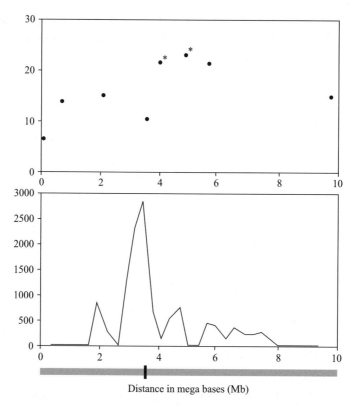

Figure 7.22   Comparison of single locus association and Bayesian multipoint LD mapping approach to localising a known susceptibility gene for breast cancer using eight microsatellite markers in 850 cases and controls. * denotes significance at the 5% level. (Adapted from Rafnar et al. 2004).

approach. Thus, multipoint LD mapping appears to be very powerful under some circumstances.

7.11.4 Further challenges

Given the amount of data being collected at present, further improvements of multipoint LD mapping methods appear justified. Having a proper coalescent prior for a heterogeneous disease by extending either the approach of Morris et al. (2002) or Larribe et al. (2002) is desirable. Integrating models for population structure, gene conversion and heterogeneous recombination rate should also be possible.

A more difficult problem appears to be efficient integration of covariates (such as age-of-onset, smoker/non-smoker, etc.) into these methods. Furthermore, the assumption that different disease mutations occur in the same position is restrictive. Methods that explicitly allow different disease mutations at different positions along the chromosome are in high demand.

For a recent example see Molitor et al. (2003). This would also include a model for the interaction of different mutations (in the same or different genes) in causing a disease.

The present optimism versus pessimism relating to the possibility of finding causative SNPs and susceptibility genes can be objectively tested for different choices of ϕ (equation (7.1)). Can the full set of causative SNPs be found for larger k? How large a sample size is needed to map the genotype to phenotype function? Since the true genotype to phenotype functions are not known, such investigations would have to be naive in this respect.

It should be stressed that the haplotype and model based LD mapping methods discussed here are not the only multilocus approaches for LD mapping. These methods were included here because they have been developed based on arguments taken from coalescent theory, and they are indeed currently very popular. However, various data mining approaches have also been developed and may outperform either of these approaches, in particular when the amount of data is limiting the use of model-based approaches due to intractability.

Recommended reading

Clayton, D. (2001), Population association, *in* D. J. Balding, M. Bishop, and C. Cannings, eds., 'Handbook of Statistical Genetics', John Wiley and Sons, pp. 519–540.

Hudson, R. R. (2001*a*), Linkage disequilibrium and recombination, *in* D. J. Balding, M. Bishop, and C. Cannings, eds., 'Handbook of Statistical Genetics', John Wiley and Sons, pp. 309–324.

Morris, A. P., Whittaker, J. C., and Balding, D. J. (2002), 'Fine-scale mapping of disease loci via shattered coalescent modeling of genealogies', *Am. J. Hum. Genet.* **70**(3), 686–707.

Strachan, T. and Read, A. P. (2003), *Human Molecular Genetics 3*, BIOS Scientific Publishers Ltd., John Wiley and Sons.

Weiss, K. M. and Terwilliger, J. D. (2000), 'How many diseases does it take to map a gene with SNPs', *Nat. Genet.* **26**(2), 151–157.

8 Human evolution

8.1 Introduction

Humans are tremendously interested in human evolution and this topic gets more attention than the evolution of any other species. This is so for several reasons. Beyond the obvious, that we are interested in our own evolution, human evolution is tied up with the interpretation and understanding of diseases with a genetic component and is influenced by human demographic and cultural history of which much is known. For other species than our own less is known about demography and culture.

Until recently, longer term human evolution (thousands to millions of years) has mainly been investigated by comparisons of dated fossils and archaeology. Such analysis has brought much understanding of the time of origin of anatomically modern humans, the major migrations and colonisation of the earth, and knowledge about demography. These studies have the advantage that they often use traits of direct interest such as brain volume, upright posture, and pathologies.

DNA sequences do not directly relate to morphological traits, but they are abundant in quantity, can be determined with accuracy and have the advantage that their evolution can be analysed by well-defined models. The smaller importance of subjective interpretation and the use of exact modelling also facilitates comparison of results among different investigations.

Although much progress has been made, there are still many contentious issues. It is unfortunate that it is so hard to obtain reliable ancestral genomic DNA on any scale, since this could settle many issues and allow a combined analysis of data from archaeology and evolutionary genetics. To get the absolute dates of events, information on the age of fossils corresponding to some internal node in the genealogy has to be used.

This chapter will address four examples where coalescent theory plays a role in understanding human evolution. The examples are:

1. Which guides do coalescent theory provide when using DNA sequences to infer our phylogenetic position among the higher primates?

2. What can coalescent theory predict about the number of genetic ancestors to the human genome and the distribution of the ancestral material on these ancestors?

3. How can we through coalescent arguments use human sequence variation to understand migration in human populations? We will address the issues of the migrations out of Africa (0.1–1.2 million years), the start of growth of the human population ($\approx$60,000 years), the relationship with the Neanderthal species and ancestral population structure.

4. Recent historical records and very large amounts of sequence data allow us to focus on the precise pedigree of special populations or in general on very recent history. How can we combine this with coalescent theory and use the information to modify coalescent models?

Obviously, due to the explosive growth in contemporary sequence data, the analyses discussed below are soon to be superseded by studies using more data, but the principles and models should remain relevant. It will be apparent that despite the often impressive amount of data, the statements that can be made about ancestral population behaviour are weak. An interesting question is the limits of this knowledge as sequence and sample number grows. Data will explode both in length (until whole genomes are reached) and number (until everybody have been sampled). This will improve reliability of obtained conclusions. Methodologically, the growth in sequence length will make it impossible to ignore recombination in analysis, and the growth in the number of sequences will make the sample size exceed the effective population size at some point.

8.2 Our phylogenetic position and ancestral population genetics

The variation observed at the population level is typically determined by events up to a few N generations back in time. Recall that $2N$ denotes the effective population size as defined in Chapter 1, which in human history generally has been much smaller than the actual physical population size. If we assume a long term effective size of $2N = 20,000$ and a generation time of 25 years, then the coalescent time scale of $2N$ generations correspond to half a million years. Assuming that the human population size has been constant at $2N = 20,000$, we can use equation (1.30) to calculate the proportion of the human genome where the coalescent time exceeds a certain number of years. The results are shown in Table 8.1 for a sample of size $n = 50$. We can see that the MRCA of about 95% of the genome sequence is less than 2 million years old. For about 0.1% a common ancestor is not found earlier than 4 million years ago, and even though this

Table 8.1   The proportion of the human genome (and number of base pairs) where the time to the MRCA is greater than t^a

Time t in $2N$ generations	Time in million years	$P(TMRCA > t)$	No of base pairs
1	0.5	0.85	$2.6 \cdot 10^9$
2	1	0.38	$1.1 \cdot 10^9$
3	1.5	0.14	$4.3 \cdot 10^8$
4	2	0.052	$1.6 \cdot 10^8$
5	2.5	0.019	$5.8 \cdot 10^7$
6	3	0.007	$2.1 \cdot 10^7$
8	4	$9.7 \cdot 10^{-4}$	2,900,765
10	5	$1.3 \cdot 10^{-4}$	392,576
12	6	$1.8 \cdot 10^{-5}$	53,129
16	8	$3.2 \cdot 10^{-7}$	973

^{a} The calculation assumes that the human population has had a constant population size of $2N = 20,000$.

sounds like a small number, it still amounts to an expected 3 million base pairs. Polymorphisms in this part of the genome may thus be sufficiently old that they are shared between humans and other primate species. Note that these estimates scale linearly with the effective population size, thus for a species with an effective size of 2,000,000, as is sometimes given as estimates for *Drosophila melanogaster*, more than 5% of the genealogies have roots deeper than 4 million years (assuming a generation time of 1/2 year).

Coalescent arguments however can be employed on much longer scales than those in Table 8.1.

First, alleles experiencing balancing selection can be much older than neutral alleles as discussed in Chapter 4. As an example, the human major histocompatibility complex (MHC) has variation dating more than 20 million years back, and shared variation with other primates has been demonstrated.

Second, if alleles are sampled in populations from extant species, this will indirectly imply sampling in the ancestral population(s) of the extant species (see Figure 8.2). This is particularly useful for recently diverged species, like human and chimpanzee, but could be used much further in time until divergence of the extant sequences has erased information about events far back in time. In addition to these two cases, it can be very useful to compare evolutionary parameter estimates obtained from within species data with estimates obtained from between species data. In short, phylogenetics is often relevant for population genetics and vice versa.

There is general agreement about our relationship to the great apes: orangutan, gorilla, and the two species of chimpanzee (see Figure 8.1).

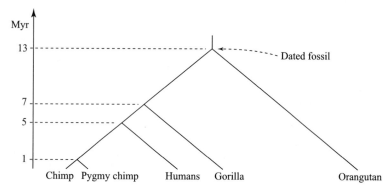

Figure 8.1   The phylogeny of the Hominoids. The two closest species are the chimpanzee species. After many decades of uncertainty, humans and chimps have emerged as slightly more related than either are to the gorilla. The dates are taken from Glazko and Nei (2003).

Besides dates of speciation, the only debatable issue has been the relationship between chimp, gorilla and humans, but it seems that chimps and humans are siblings in this triad. These dates have become increasingly reliable in the last decade due to better models of sequence evolution and much more sequence data. Often the date of the common ancestor of human and orangutan is taken to be 13 million years and then used to date all other points in the phylogeny.

In reconstructing the phylogeny of distantly related species, the polymorphisms of the speciating population can be ignored, since the time back to a common ancestor for two alleles within a population is usually small relative to the internal branches in the species phylogeny. But the bifurcation into humans and chimp occurred so soon after the splitting-off of the gorilla lineage, that polymorphism in those ancestral populations were of importance and the relationship between the three species depends on which gene is used for analysis. Chen and Li (2001) investigated fifty-three intergenic regions and found that thirty-one supported human and chimpanzee as sister species, ten human and gorilla as sister species, and twelve gorilla and chimp as sister species. Together, the fifty-three genes strongly support a phylogeny in which human and chimps are evolutionary closest. Chen and Li (2001), Takahata et al. (2001) and Yang (2002a) have tried to estimate the population size of the MRCA to humans and chimpanzees by exploiting the fact that gene trees often differ from the species tree (21/53 = 40% of the time in the study discussed above). If the ancestral population size and times of MRCA of the (human, chimpanzee) clade and of the (human, chimpanzee, gorilla) clade were known, the probabilities of different topologies for a gene could be calculated easily. At the time of speciation, T_2, there will be two distinct alleles, a chimpanzee and a human allele (see Figure 8.2). If these two alleles do not find a common ancestor in the time period of $T_1 - T_2$, there will be three ancestral alleles in the ancestral population at time T_1. If the two alleles that are ancestral to human and

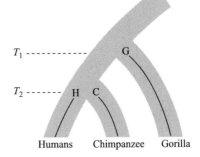

Figure 8.2   The size of the population ancestral to humans and chimpanzee. Estimating this can be tied to the probability that the gene tree is the same as the species tree. What is important is the ratio between N and $T_1 - T_2$. A small $N/(T_1 - T_2)$ will make identity between gene tree and species tree highly probable.

chimpanzee alleles do not find a common ancestor with each other before they find a common ancestor with the ancestral gorilla allele, then the allele phylogeny and species phylogeny will differ. If the generation time is G and $2N$ constant, then the probability of this is

$$p = \frac{2}{3} e^{-\frac{T_1 - T_2}{2NG}} \tag{8.1}$$

or equivalently

$$N = -\frac{T_1 - T_2}{2G \ln(\frac{3}{2} p)}, \tag{8.2}$$

which follows from the coalescent assumption of an exponential waiting time until a MRCA. There should be equal chance for the two wrong phylogenies, as also seen in the above example where the two wrong phylogenies were observed in ten and eleven cases. If for a set of genes one phylogeny is in convincing majority, the frequency of the two remaining could be used as p. If $T_1 - T_2$ is assumed known, then N can be calculated. For example, assume $T_1 - T_2$ to be 1.5 million years, generation time 20 years, and p to be 0.1, then N could be estimated to 39,700 individuals. Chen and Li (2001) and Takahata et al. (2001) reconstructs the phylogeny and uses the above equation to estimate $N/(T_1 - T_2)$ directly. Yang performs a more complete analysis that models the data directly. In principle the latter approach is superior, since it does not rely on having the correct phylogenies. Yang estimates the effective population size to be in the range 12,000–27,000, considerably lower than the estimates of Chen and Li (2001) and Takahata et al. (2001).

8.2.1   The number of genetic ancestors to a genome

Here we will elaborate on the genealogical process applied to the human genome. Above, we discussed the distribution of TMRCA of the genome given a model of population size. However, different positions on the chromosome are correlated as determined by the coalescent.

Since the coalescent with recombination generates a family of trees, it is clear that the expected time (maximum over all) until all positions have found a common ancestor must be an increasing function of sequence length. No expression exists for this function, but it can be estimated by simulation as shown in Figure 8.3 for ten sequences. Note that within a segment of length $\rho/2 = 64$, which is on the order of 1 Mb in the human genome, we expect the highest tree has a TMRCA of $7.5N$ generations, that is, 3.75 million years. Also shown in Figure 8.3 is the earliest time until a position has found a common ancestor (minimum over all) which is a decreasing function of sequence length. Extending this to the complete genome to see how old some ancestral segments could be, would of course be of interest.

Another question of interest is on how many chromosomes among the set of ancestral chromosomes, the ancestral material to a sample will eventually be distributed. For the Y chromosome and mitochondrion the absence of recombination implies that there is only one ancestor for each. The individuals having these sequences have been called 'Adam' and 'Eve' respectively.

For all the autosomal chromosomes, recombination distributes the genetic material on more than one ancestor. One way of studying the process of distributing genetic material on ancestors is to follow one sequence back in time until the process of recombination and coalescence has reached an equilibrium. This has been done by Wiuf and Hein (1997).

The left of Figure 8.4 shows the outcome of one such process. The ancestral chromosomes have been labelled in the figure by scanning the sequences carrying ancestral material from left to right using integers starting with 1 and using an integer 1 higher than any previous integer each time a new

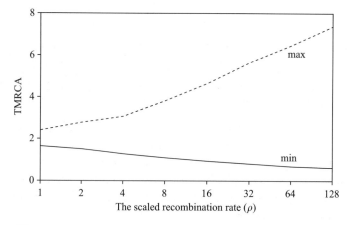

Figure 8.3   The average minimum and maximum times until all positions in ten sequences have found their MRCA as function of sequence length.

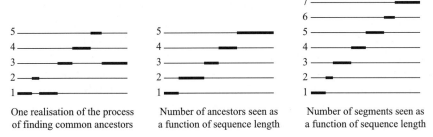

One realisation of the process
of finding common ancestors

Number of ancestors seen as
a function of sequence length

Number of segments seen as
a function of sequence length

Figure 8.4   One realisation of the process of finding ancestors to a single chromosome or sequence. In all three illustrations, the sequences are traversed from left to right. In the left illustration, the ancestral material at the leftmost position has been traced back to a chromosome that will be named '1'. Moving right, '1' will have a contiguous segment with ancestral material after which the ancestral material will be found on another chromosome that will be called '2'. After '2's segment the next segment happens to be on '1' again. The next segment is found on yet another chromosome thus called '3', etc. The middle illustration shows how many ancestors there are, and the right illustration shows how many segments have been observed moving left to right.

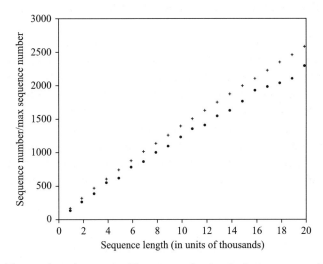

Figure 8.5   The number of ancestors. The top set of points is the average number of ancestors, which for $\rho = 20{,}000$ would be about 2,600. The average number of the chromosome having the rightmost segment is about 2,300. If this is extrapolated to the full length of chromosome 1, the expected number of ancestral segments is 52,000 carried on 6,800 sequences. (Wiuf and Hein 1997.)

sequence is encountered. This chromosome scanning will encounter small islands of ancestral material. Two questions are: How many islands will be encountered, and on how many sequences are they distributed? These two quantities are shown as a function of sequence length in the right and middle part of Figure 8.4. We will map the mean of these quantities as a function of ρ by simulation.

Figure 8.5 shows the expected number of ancestral sequences as a function of ρ. Exemplified by chromosome 1 with a population size of

$2N = 20,000$ and uniform recombination rate: Assume $\rho = 1$ corresponds to 5 kb, then chromosome 1 has a length of $\rho = 52,000$. Figure 8.5 also shows the average label of the last sequence. This is bound to be less than the number of ancestors, but happens to be only slightly smaller. The average number of segments is $1 + \rho$—one of the few quantities that can be calculated. The number of the individual carrying the last ancestral piece as a function of ρ is also interesting. In the very left realisation in Figure 8.4, the last ancestral segment (seventh) is carried by sequence 3, which also carries the fourth ancestral segment. It is seen in Figure 8.5 that the average number of these sequences is quite close to the number of sequences carrying ancestral material. How is this to be interpreted? As the chromosomes are traversed left to right, tracing ancestral material, it is possible to jump back to earlier chromosomes, but it is unlikely to jump back to chromosomes with a much lower number. So the process jumps to new chromosomes with lower and lower rate (the curve is sublinear), but it mainly jumps back to recently visited chromosomes. Although this process is non-Markovian, the dependence on the realisation further back is weaker.

Figure 8.6 shows part of a randomly chosen ancestor in these simulations. The structure of such an ancestor (autosomal Adam or Eve) is also interesting: Ancestral material comes in little batteries, which is in line with the comments to Figure 8.5, that this process is quite local. It visits some chromosomes a number of times within a small region and then leaves those chromosomes most likely for good. This ancestor also shows that the amount of trapped material is quite substantial: All material from 6894 to 8352, except the nine little ancestral segments, is non-ancestral. So although the ancestral material is close together relative to a uniform distribution along the chromosome, it is still out-factored by non-ancestral material by a factor of about 150. In simulations this implies that most recombination events will be in trapped non-ancestral material.

If these quantities are calculated for all chromosomes, it gives about 86,000 ancestors. How these quantities behave under a more realistic model remains to be explored, but most of our ancestral 'Adams' and 'Eves' are not carriers of the mitochondrion or Y chromosome, but of little pieces of ancestral material to the autosomes.

Yet another quantity of interest is how the amount of non-ancestral bridges in the graph representing these genealogies grows as a function of sequence length. Unfortunately, it grows enormously. At first the trapped ancestral material is zero but quickly it grows to an amount considerably larger than the actual ancestral material of interest. The amount then levels off to a lower value as the ancestral material finds common ancestors and its total amount reduces. It will not tend towards zero as the ancestral material to only one sequence will also trap non-ancestral material as shown in Figure 8.6. It is the large amount of this trapped material that causes a slowdown in simulation algorithms. Recombinations within this material

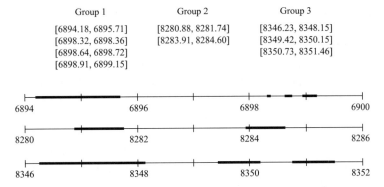

Figure 8.6 A random ancestor and its distribution of ancestral segments. An obvious and intuitive characteristic is apparent from this example: An ancestor has a series of segments and these are placed in small groups. In this case nine segments in three groups and the groups are also close on the whole chromosome. This is because close segments are more stable than distantly placed segments. The expected duration of a physical linkage of two segments is inversely proportional to the distance between them, since they are disrupted by a recombination between them. With the chosen parameters the average length of a segment (one unit on the illustration) is 5 kb (Wiuf and Hein 1997).

Table 8.2 Expected number of different types of events and genealogies for complete genomes

Sample size	Recombinations	No. of tree changes	No. of topology changes
2	80,000	53,200	0
10	220,000	154,000	66,000

and creation of trapped material by coalescent events must be traced and simulation algorithms spend most of their time doing this.

The number of distinct genealogies for a set of complete chromosomes sampled from the human population determines for example the limits to LD mapping (Chapter 7). To get a rough guess of how such a number of genealogies would look, let us translate the considerations from previous paragraphs to the idealised human population of 10,000 and with lengths as are known for the human chromosomes. Several questions are of interest; we will only consider one: How many different trees would be encountered if all sampled chromosomes are scanned?

If we use the quantities from Table 8.2 and enquire about the genealogical history of ten genes, the expected number of recombinations, the number of times a tree change occurred and the expected number of a tree topology changes occurred, can be calculated.

A little care has to be given in moving from one chromosome to the next. It does not make sense to speak of recombination between the last position of chromosome 1 and the first position of chromosome 2, but there is a

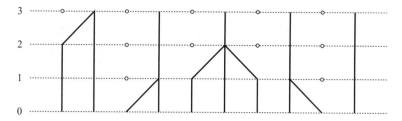

Figure 8.7   The whole population consisting of ten genes was sampled and observed for four generations. In the transition from generation 0 to 1 two simultaneous coalescent events occurred—(3, 4) and (8, 9). In the next transition from generation 1 to 2, a triple coalescent event occurred.

small possibility these positions have the same tree topologies. However, these probabilities are small and in answering these questions it makes little difference if the whole genome was regarded as one long chromosome.

These calculations should be taken with some caution. First, they are based on the simplest conceivable population model. Second, the number of ancestral sequences is so high that the assumption that multiple and simultaneous coalescence events can be ruled out, is not warranted. So the intensity of coalescence is most likely higher, making the numbers obtained upper bounds on the true number. In investigations involving whole chromosomes or genomes the assumption that the sample size is small compared to the population does not hold. If a very large number of genes (for instance the whole population) is sampled, this problem can still be ignored, since the coalescent process will quickly bring the large number down and then the assumption does hold. However, for genomic analysis the sample size will continually be high due to a large intensity of recombination. The assumption of small sample size precludes simultaneous coalescent events and coalescent events involving more than two lineages. If the standard theory is applied without taking this into account, then the intensity of coalescent events is underestimated.

Figure 8.7 illustrates the occurrence of simultaneous and coalescent events involving many lineages. Pitman (1999) and Schweinsberg (2000) have considered the combinatorics of such events.

8.3   **Human migrations and population structure**

Since the human versus chimpanzee speciation, a series of changes has occurred eventually leading to modern humans. Upright walking evolved in our ancestors 2–3 million years ago in Africa. There was an exodus about 0.8–1.3 million years ago leading to *Homo sapiens* fossils in Europe and

Figure 8.8   Twice out of Africa with total replacement (left) and with partial replacement models (right). These models have different consequences for the probability of ancestry of genes from the three regions. In a total replacement scenario, genes will find common ancestors not much further back than the second (most recent) exodus in the ancestral population that produced the emigrants. In the partial replacement model, genes could have resided in the populations in Asia and Europe that had been there since the first Exodus, and a common ancestor would be more than a million years old.

Asia. An anatomically fully modern man was first present 130–150,000 years ago and then spread out to Europe and Asia.

It has been debated if the second exodus fully replaced the first exodus (the total replacement hypothesis) or if there was coexistence and interbreeding. Idealised versions of the contending hypotheses are shown in Figure 8.8. The total replacement hypothesis has been favoured by population geneticists, while the partial replacement hypothesis has been favoured by a few paleontologists from morphological arguments.

The two scenarios assign different probabilities to gene genealogies. A complete displacement will add more weight to genealogies with a common ancestor for genes taken from the different groups more recent than 200,000 years ago. A partial replacement model allows some genes to have an ancestry at the first, some at the second exodus. Takahata et al. (2001) has again tested this and varied a set of the relevant parameters, such as the sizes of the founding populations and allowing for locally and globally adaptive variants. The data was not sufficient at the time for a final conclusion, but pointed towards an almost complete replacement, since most genes genealogies are compatible with a MRCA time in Africa.

Based on allele frequencies Cavalli-Sforza and collaborators have used the techniques of principal components projected on a map of Europe in an attempt to trace migrations and demographic changes within the last 10–20,000 years. Figure 8.9 shows the most famous of these projections that has been interpreted as the demographic replacement of hunter-gathers due to the advance of agriculture. Thus interpreted, this illustration would favour the theory that agriculture was coupled to a population expansion in contrast to agriculture being transmitted as a culture. These illustrations have been severely criticised (for instance Sokal et al. 1999) and the gradients described as artifacts produced by a smoothing procedure used by

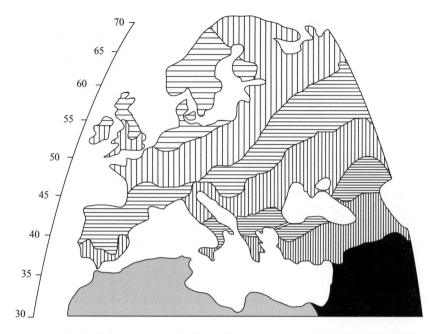

Figure 8.9   The gradient corresponding to the first principal component is projected upon the map of Europe. This gradient is interpreted as the population replacement of hunter-gatherers by agriculturalists. Similar illustrations exist for the second to fifth principal component that also have been given historical interpretations. (Adapted from Cavalli-Sforza 2001.)

Cavalli-Sforza and collegues. As allele frequency data increasingly will be supplemented by sequence data, the natural model for analysing such historical migrations would be a coalescent model on a suitably scaled plane with and without drift (see Chapter 4). Such analysis could also answer questions about the possibility of dating ancient migration events.

8.3.1   Our relationship to the Neanderthaler

A closely related issue is the question of our relationship to the Neanderthal species, which is still not fully resolved. The first Neanderthal fossils are more than 100,000 years old and the most recent fossil is about 28,000 years old. The Neanderthal was morphologically distinct from *H. sapiens*, but the time back to a common ancestor, the possibility of interbreeding and the effective population size of the Neanderthal are key unsolved questions. Getting enough sequences from Neanderthal fossils could eventually resolve these issues. However, it will take more than what is presently available. Mitochondrial sequences are the only sequences obtainable in reasonable amounts with present technology. The first mitochondrial sequence was published in 1997 by Svante Paabo's group together

with an analysis (Krings et al. 1997). The analysis contained a very interesting error that was pointed out by Nordborg (1998). Paabo's group
sequenced one Neanderthal segment from the hypervariable region of
the mitochondria and analysed it together with 986 extant *H. sapiens*
sequences. The Neanderthal sequence came out as an outgroup to the
986 humans. This seems solid evidence that Neanderthal is an outgroup
to *H. sapiens*, except for one important fact: The Neanderthal sequence is
30–100,000 years old, while the human sequences were modern. The fair
comparison is to let the 986 human sequences coalesce corresponding to
these years. For this much smaller number of ancestors one should ask if it
is unlikely that the Neanderthal sequence comes out as an outgroup even if
it was part of the *H. sapiens* population. Nordborg performed these fairer
comparisons under a set of scenarios including growth and merging of the
two populations at a specified date. In Table 8.3, the conclusions from the
simplest scenario is shown (see also Chapter 3). Generation time was set
to 20 years and the female effective population size 3400. The effects of
letting the human sequences coalesce for 30–100,000 years are startling,
but still should not surprise. The number of ancestors to the present 986
sequences is seriously reduced and the probability that the ancestor appears
as an outgroup is high, even if the underlying breeding model assumes that
the Neanderthals and humans were part of the same population.

The single Neanderthal sequence has since been augmented by more
sequences and although Neanderthals still come out as a separate group,
the probability of this occurring is still not sufficiently strong to justify the
present strong statements about genetic differences between Neanderthals
and *H. sapiens*.

Table 8.3 Three key quantities under constant and
exponential population scenario[a]

	Constant	Growth
Number of ancestors	4.86	782
P(observed topology)	0.085	$2.3 \cdot 10^{-6}$
$P(T_c \geq 4T_t)$	0.0063	$3.7 \cdot 10^{-8}$

[a] The expected number of ancestors to the present
H. sapiens sample of size 986 (upper row), the probability
of observing a topology with the Neanderthaler as an
outgroup even if he was a member of the *H. sapiens*
population (middle row) and the probability that the time,
T_c, back to a MRCA of both the Neanderthaler and
H. sapiens, is more than four times larger than the time,
T_t, back to the MCRA for *H. sapiens* excluding the
Neanderthaler (lower row). Numbers are from Nordborg
(1998).

Mitochondrial DNA is most likely not going to resolve these issues, since they will only constitute observations from one genealogy generated from the historical reproduction structure of *H. sapiens* and Neanderthals. Autosomal sequence data will be much more valuable in the study of population structure of archaic Neanderthals and humans, but at present is much harder to acquire.

8.3.2 Population growth

It is an established fact that there has been a tremendous growth in population size starting with the Agricultural Revolution 12,000 years ago, a global population of an estimated 50–200 million individuals increased to the present 6 billion individuals. Inference on population structure and growth in the period further back has been attempted. Wall and Przeworski (2000) have tested diverse growth scenarios from sequence data using a variety of coalescent simulations. One of their more surprising observations has been that long term growth is not as easily detectable as expected from knowledge of demography and archaeology. A partial explanation is that much of the growth has been very recent (200 years), however, disregarding this, there has still been a significant long term growth.

Any specified scenario (population constancy, growth, bottlenecks, mutation rate, etc.) will define a distribution of the nucleotides for a site in a sample from a population. The vast majority of sites will be non-segregating, extremely few sites have more than two nucleotide states in a position and the rest will have two nucleotides observed in that site. Perfect knowledge of this distribution would allow strong inference about the population scenario. If all sites were independent and many sites were observed, this would give good, but not perfect, information about this distribution.

8.3.3 Structure within global modern human populations

The present era is dominated by growth and unprecedented migration and the genealogical information in a person's geographical position is less than earlier. The last centuries and to a lesser degree the last 12,000 years has experienced major growth. This creates certain problems when analysing sequence data from the present. The value of sequence data, beyond forensics, is mainly in its use in making statements about events further back in time. The more recent events are often historically documented, but could eventually be traced using sufficiently large amounts of sequence data. In principle appropriate coalescent models and data should be able to handle questions about population structure of ancient time periods. But at present both data and efficient simulation algorithms are limited.

One of the most effective approaches to this at present is the non-coalescent based method implemented in the program STRUCTURE

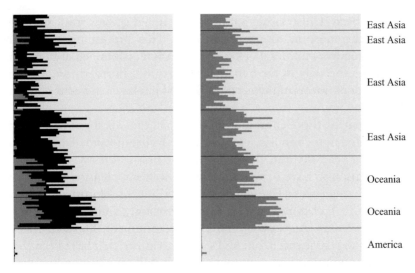

East Asia

East Asia

East Asia

East Asia

Oceania

Oceania

America

Figure 8.10   The x-axis shows the probability of ancestry of locus, individual. The y-axis is the geographical origin of the individual. The analysis of Rosenberg et al. (2002) involved more ancestral populations and more loci than shown here, where only $K = 2$ (right) and 3 (left) are shown. When $K = 2$ the individuals will be explained by two unnamed populations—indicated by two shades of grey. In going to $K = 3$ the individuals from Oceania will be composed by three populations (indicated by three shades of grey), while some individuals will not have any contribution from the third population. (From Rosenberg et al. 2002, with permission from publisher.)

(Pritchard et al. 2000). Given a set of individuals and their genotypes at a set of loci and a preset number, K, of fundamental populations, STRUCTURE will estimate founder population genotype frequencies at the loci for the K populations. Each individual will at each locus be assigned probabilities for belonging to each of the K populations.

A recent application of the STRUCTURE program is illustrated in Figure 8.10 in the analysis of Rosenberg et al. (2002). A total of 377 autosomal microsatellite loci from fifty-two populations were used. Using this approach, data can be preprocessed for further analysis using an explicit coalescent model, instead of explicit modelling of population structure in a coalescent model.

Although this is not coalescent based analysis, it is very useful in defining structure in a population that has become hard to define geographically, due to massive recent migrations.

8.3.4   Specific histories

The analysis of future genomic data will be increasingly reliant on inference methods involving the coalescent with recombination. At present this is avoided, since the sampled sequences are regarded as independent genealogical outcomes of the same unobservable population structure.

There has been a historical trend in data use in going from mitochondria to Y chromosome to autosomal data. This is quite understandable. The mitochondria mutates quickly and variation is obtained even for short sequences—especially in the hypervariable region. Furthermore, the mitochondrion does not recombine, so traditional phylogenetic methods can be applied.

Evolutionarily, the Y chromosome is the paternal analogue to the maternally inherited mitochondria. However, the Y chromosome DNA mutates much slower than the mitochondria and the first studies were in the frustrating situation to have to wrench phylogenetic information from a series of identical sequences (Hammer et al. 1997). As sequencing methods improved, the Y chromosome became increasingly valuable (Jobling and Tyler-Smith 2003). An issue at present is ascertainment. The whole Y chromosome is not resequenced, but SNPs are determined that have been chosen by certain population investigations and this will give bias, when applied to other populations.

Contrasting results obtained from Y chromosomes and mitochondria can reflect differences in reproduction among the sexes. Seielstad et al. (1998) and later Oota et al. (2001) contrasted pairwise divergence in the Y chromosomes and mitochondria in three patrilocal (the wife moves when married) and three matrilocal (the man moves when married) societies and there was a clear confirmation that matrilocal societies would have little within society mitochondrial variation, relative to patrilocal societies. And the trend was reversed for Y chromosome data.

8.3.5   Empirical pedigrees and the coalescent

Coalescent theory makes probability statements about the genealogy relating a set of sequences from a population. The relationship among individuals—pedigrees—has a different structure from both a phylogeny and the ancestral recombination graph, since individuals always have two parents (mother and father). Pedigrees are known in great detail for certain groups. The size of even the largest of these pedigrees is vanishing relative to the pedigree of the whole human population, but nevertheless the pedigrees represent information about the genealogical process that coalescent theory tries to model. Most individuals will have genealogical information about his/her ancestry in knowing parents and the identity of some ancestors further back in time. A celebrated example of a large pedigree is the Icelandic pedigree that has been collected by the company Decode and goes back a millennium and involves 260,000 people (Helgason et al. 2003). Other large pedigrees are known for Mormons, Amish, and part of the Quebec French (Austerlitz and Heyer 1998). The largest and deepest pedigrees registered are for Chinese clans (Zhao 2001). In England a celebrated example is the British Peer Pedigree (Westendorp and Kirkwood 1998). Many of these pedigrees are associated with biases and errors, limiting their

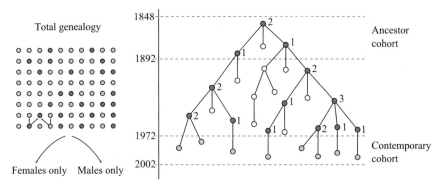

Figure 8.11   From the genealogy of the population, the paternal and the maternal genealogies were extracted. Of these pedigrees only individuals with survivors born after 1972 were included. This gives two realisations of a coalescent process starting with extant individuals and going towards their common ancestor. In the left panel the light grey nodes are females and dark grey males. Only keeping light grey (or dark grey) nodes and edges connecting these would extract a female genealogy (or male genealogy). In the right panel, light grey nodes represent the cohort of individuals born after 1972 and before 2002 (only males or only females). The dark grey nodes are the ancestors to the present cohort in the male or female genealogy. The white nodes are descendants of the ancestors, who did not have any descendants in the present cohort. (Adapted from Helgason et al. 2003, with permission from the publisher.)

use for demographic purposes. Such biases can be social, arbitrary cutoffs in how far family ties are traced back and geographically related variation in reliability of asserted family relationships.

If the pedigree of an individual is known, the genetic material has a genealogy that is contained within the pedigree. In the case where the genetic material is not subject to recombination a phylogeny within the pedigree would be found, however in presence of recombination an ARG would be needed to describe the history of the genetic material. Knowledge of the pedigree thus restricts the ancestors a sequence can choose. It is an interesting, but seemingly unexplored, question to ask how useful these constraints could be in association mapping (Chapter 7).

The work of Helgason et al. (2003) used the Decode database with complete genealogical information far back in time. To focus on specific questions, only the paternal parts and maternal parts were extracted, giving separate genealogical relationship of all the women and all the men, respectively. This can be interpreted as making the genealogy of the mitochondria of all woman and the Y chromosome of all men (see Figure 8.11). This reduction to a male and a female genealogy is a strong reduction and many additional questions could be posed using the full pedigree. Within the male and female genealogies only individuals with descendants born after 1972 were kept thus removing a series of males and females that did not leave progeny in a direct male, respectively female, direct line born after 1972 (Figure 8.12).

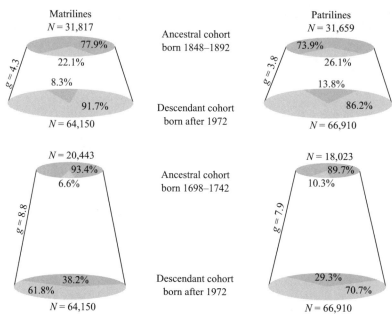

Figure 8.12 Tracing male and female lineages back to two earlier periods. The right part of the figure defines paternal contributions to the present from the two historical periods and the left the maternal. The average number of generations from the historical epochs to the present is *g*, next to each case and again illustrates the shorter female generation time. The numbers at the lower right cone is the number of individuals in the 1698–1742 population: 10.3% have sent Y chromosomes to the present population, while 89.7% have not. In the bottom of that cone, there are 66,990 males born after 1972, 70.7% can have their Y chromosomes traced back to the 1698–1742 population, while 29.3% cannot. The 29.3% could be due to immigration. (Adapted from Helgason et al. 2003, with permission from the publisher.)

Within these two genealogies two ancestral populations (cohorts) were defined by time periods—1698 to 1742 and 1848 to 1892 (Figure 8.13). These 45-year periods are so long that the extant individuals should have an ancestor in this period, if the ancestral line was continued in Iceland. A shorter period like 25 years could be escaped by having one ancestor born after the period and the previous ancestor born before. It is seen that a small fraction back in the historical period is responsible for a large fraction of the present population and that a large fraction do not have any descendants in the present population. A certain fraction do not have an ancestor in the historical period and since everybody surely had ancestors, this must be due to immigration (or possibly a little 'cohort skipping'—having an ancestor born after the earlier epoch (for instance 1893), while that ancestor's parent was before that epoch (for instance 1846)). Going backwards in time, the ancestor must then be found outside Iceland. For the 1848–1892 cohort 22.1% were responsible for the

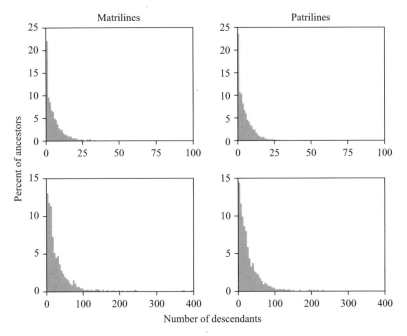

Figure 8.13 Illustration of the number of direct male descendants (right part) of males from the historical epochs and similarly for females (left part). The upper part is for the 1848–1892 period and the lower part from the 1698–1742 period. Individuals from the epoch further back will contribute a larger number of offspring to the present cohort. (Adapted from Helgason et al. 2003, with permission from the publisher.)

present cohort and this number was naturally smaller (6.6%) for the 1698–1742 cohort. Going sufficiently far back would have defined an 'Eve' for the Icelandic population, that most likely existed much before the colonisation of Iceland and therefore also outside Iceland. For females again 8.3% in the present cohort cannot be traced back to the 1842–1892 cohort.

A series of interesting conclusions can be drawn from this study (not all entirely new):

1. The maternal and paternal coalescent processes run at different generation times, since female generation times are about 0.9 times shorter than male generation times.

2. There is greater variation in offspring numbers among women than men again accelerating the coalescent process in the female lines.

3. There is a correlation in offspring numbers between parent and offspring and this is stronger in females than males. This runs counter to the Markov property assumed in the coalescent process. The continuous time approximation to the discrete coalescent can still be valid if this correlation dies out sufficiently quickly. Donnelly and Marjoram (1989) have analysed models allowing for this.

Obviously, these are important lessons if coalescent modelling is proceeding towards a more detailed modelling of more recent human population. In the analysis of variation data created over longer time periods, the coalescent approximations about generations are less problematic.

8.3.6   Other genealogical issues

8.3.6.1   *Tracing both parents*

Chang (1999) has investigated the growth of the number of ancestors as a function of generations in an idealised constant sized population where each individual chooses his/her parents independently of others. Obviously, any individual has two parents and most have four grandparents and eight great grandparents, etc. If this trend is continued for k generations back an individual should have 2^k ancestors. Obviously the true number would start to be smaller than this as some ancestors (repetitions) could be reached by multiple paths as the number of ancestors grows large (see Figure 8.14).

Chang proved that the expected time until the first occurrence of an individual that is the ancestor of everybody is $\log_2(N)$ implying that the growth process is slowed significantly by the appearance of repetitions.

This result is related to the existence of one universal ancestor: There is a time when everybody is an ancestor of the whole extant population or of no one in the extant population. Obviously, this time will be further back in time than $\log_2(N)$. Chang proved that the expected waiting time until everybody had become the ancestor of the whole extant population or of no one was less than $1.77\log_2(N)$. The fraction of universal ancestors will be close to 80% and the fraction that is ancestors of no one in the present population is close to 20%.

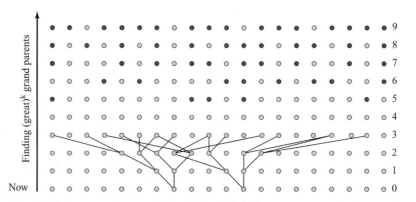

Figure 8.14   Tracing the ancestors of an individual will define a fast growing population for each individual. These populations will start to overlap and the question is when does the first individual appear that is in all these ancestor populations? At some point further back in time each individual is either the ancestor of everyone or of no one. Dark grey nodes are ancestors of everybody, while light grey are not.

Paths

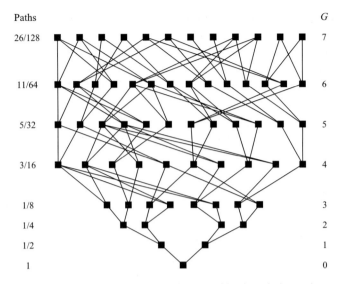

Paths		G
26/128		7
11/64		6
5/32		5
3/16		4
1/8		3
1/4		2
1/2		1
1		0

Figure 8.15 Here the ancestry of an individual is traced back and obviously an ancestor can have several lines leading up to it. The numbers to the right are the generation numbers, starting with the sampled individual. The numbers to the left are the number of paths from the sampled individual leading to the left most ancestor in that row, divided by the number of paths leading to that generation. In the top line, the 26/128 means that out of 128 (2^7), there are 26 paths to the left most individual in that row. This fraction is also the expected contribution of that ancestor to the sampled individual. The number of paths doubles each generation further back in time. (Adapted from Derrida et al. 2000, with permission from the publisher.)

Derrida et al. (2000) have independently undertaken similar investigations and also went beyond the 'yes' or 'no' question addressed by Chang in asking how many times an individual was an ancestor. Radiating from a given individual to ancestors k generations back will be 2^k paths and some ancestors will obviously be reached by many paths (see Figure 8.15). The fraction of the 2^k possible paths leading to a given ancestor is the expected fraction of genetic material that an ancestor contributes to the present individual.

Unlike the coalescent process, the ancestor process has not been extended to more realistic scenarios concerning geography, population history and selection, so illustrations using this theory are bound to be naive. However, for the purpose of illustration, assume Denmark's population is obeying the assumptions: 6 million people going back indefinitely, without social structure and a generation time of 25 years. To be almost certain an individual had either no descendants in the present or was the ancestor of all in the present we would have to go back $1.77 \log_2 (6,000,000) \cdot 25$ years $= 966$ years. Since King Gorm (the first Danish King—d. 958) has descendants in the present day population he is obviously an ancestor of all in the kingdom.

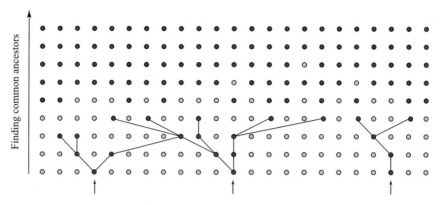

Figure 8.16 The ancestor process will randomly pick pairs of parents and the number of ancestors will typically grow very quickly in the beginning. The set of ancestors of two individuals will probably start to overlap if going sufficiently far back in time. At some point an individual could exist that was the ancestor to everybody in the extant population. Ancestral individuals whose lineages have gone extinct will be the ancestor of nobody in the extant population (or all future generations for that matter). The ancestor process and the coalescent process can be viewed as special cases of a more general process, if it was possible for individuals to have one or two ancestors with probability p and $1 - p$ respectively. Black dots indicate ancestors of the three sampled individuals.

Wiuf and Hein (1999a) investigated how this process changed, if an individual had a random variable number of parents larger than 1. For plants, parent number between one and two can be interpreted as the degree of selfing (see Figure 8.16). Species with parent numbers higher than two are not known. A motivation for this is also to try to relate the ancestor process to the coalescent process, by sampling a set of individuals and then letting the probability of only having one parent p converge to one. The coalescent process happens on a scale of N, while the ancestor process happens on a scale of $\log_2(N)$. The generalised ancestor process occurs on a scale of $\log_p(N)$ when p is larger than one and jumps to the coalescent scale for $p = 1$.

8.3.7   Tracing genetic material within the parent genealogy

The coalescent process traces genetic material while the ancestor process traces ancestors. The first shrinks and the second grows going back in time (see Figure 8.17). It is of interest to ask a puzzling question that combines the genetic and the ancestry perspective: How far back in time is the first ancestor that did not manage to contribute to our genome?

This was addressed by Kevin Donnelly (1983) and the expected appearance of this ancestor is typically not very far back in time (see Figure 8.18). If there was no recombination and restricting ourselves to the autosomal chromosomes, each individual would have forty-four chromosomes to give

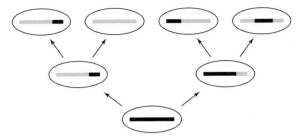

Figure 8.17 The genealogy of genetic material (the ARG) within the pedigree of ancestors. The traced segment had two ancestors a generation back, but only three two generations back. Without recombination there would be one ancestor in all generations for a single chromosome. In the first few generations, there will be no repetitions (ancestors that can be reached several ways) in the pedigree and thus no coalescent events in the ARG.

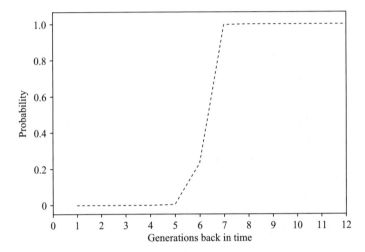

Figure 8.18 Waiting for genetically non-contributing ancestors. This curve is the probabilities (obtained from simulations) that there exists an ancestor that did not contribute to the present individual. This probability is obtained under the assumption that the number of ancestors grows as a power of two. (Due to Yun Song, personal communication).

the two parents and in the first step of this process each of the two parents will get twenty-two chromosomes. In the next generation for instance the mother could pass these twenty-two chromosomes on to the two grandparents randomly. Since six generations back there are sixty-four ancestors and there are only forty-four chromosomes, so without recombination already after six generations there would have to be non-contributing ancestors. The presence of recombination increases this number a bit, but already at seven generations back there is more than 0.5 probability of a non-contributing ancestor.

Recommended reading

Cavalli-Sforza, L.-L. (2001), *Genes, People and Language*, Penguin.

Helgason, A., Hrafnkelsson, B., Gulcher, J. R., Ward, R., and Stefansson, K. (2003), 'A populationwide coalescent analysis of Icelandic matrilineal and patrilineal genealogies: evidence for a faster evolutionary rate of mtDNA lineages than Y chromosomes', *Am. J. Hum. Genet.* 72(6), 1370–1388.

Jobling, M. A. Hurles, M. E. and Tyler-Smith, C. (2004), *Human Evolutionary Genetics*, Garland Science.

Kammerle, K. (1989), 'Looking forward and backwards in a bisexual moran model', *J. Appl. Prob.* 27, 880–885.

Wiuf, C. and Hein, J. (1997), 'On the number of ancestors to a DNA sequence', *Genetics* 147(3), 1459–1468.

Appendix: Web based tools

A number of web based tools have been developed in connection with this book and are available through www.coalescent.dk. The tools provide visualisation of the coalescent process, simulation of sequences under the coalescent, and coalescent based analysis of sequence data. It is our experience that usage of these tools may help one to build a good intuition about coalescent theory.

A.1 Wright–Fisher and Hudson animators

These two tools illustrate properties of the discrete and continuous time coalescent, respectively. In the Wright–Fisher animator (screen shot in Figure A.1) the transmission of genes can be followed generation by generation. In the Hudson animator, coalescent genealogies are gradually built-up under a range of model assumptions, for example, population growth,

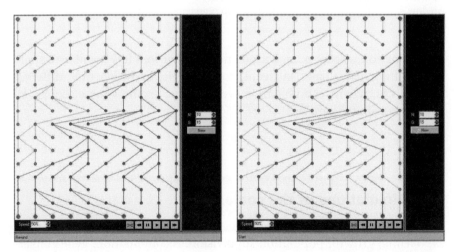

Figure A.1 Screen shots from the Wright–Fisher animator. Ten genes are followed in fifteen generations: (a) forward in time: the descendants left by two different initial genes: (b) backwards in time: the ancestors to a sample of three genes, including the coalescent events.

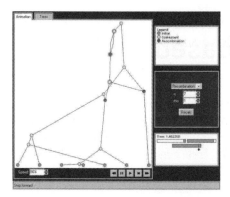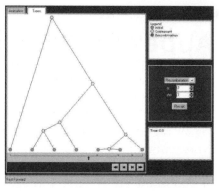

Figure A.2   Screen shots from the Hudson animator with recombination. A sample of $n = 7$ genes with a recombination rate of $\rho = 1$ is depicted.

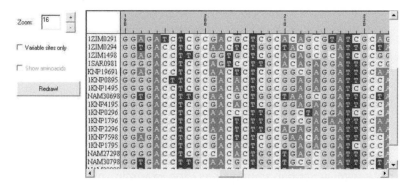

Figure A.3   Screen shot from the DataAnalyser application, showing the alignment window with designation of segregating sites to synonymous or non-synonymous.

recombination, selection, and population subdivision. It is possible to study the resulting genealogies in detail, for example, at which time different events occur, where recombination events occur, and how these change the tree (screen shot Figure A.2).

A.2   Simulations of the coalescent

Simulation of sequences can be done under the coalescent process with exponential growth, population subdivision, recombination, gene conversion, or any combination of these. There is a choice between the finite sites model or the infinite sites model of mutation.

A.3   Analysis under the coalescent

The DataAnalyser tool (screen shot, Figure A.3) allows the user to upload an aligned sequence data set. This could be a real or a simulated data set or one of the example data sets used in the book. A number of simple statistics can be calculated on the data set, including the number of segregating sites, the average number of pairwise differences, Tajima's D, the ratio of non-synonymous to synonymous sites (dn/ds) by the method of Nei and Gojobori (1986), the ratio of transitions to transversions (ts/tv), and simple estimators of the recombination rate.

Bibliography

Abecasis, G. R. and Cookson, W. O. (2000), 'GOLD–graphical overview of linkage disequilibrium', *Bioinformatics* **16**(2), 182–183.

Abecasis, G. R., Noguchi, E., Heinzmann, A., Traherne, J. A., Bhattacharyya, S., Leaves, N. I., Anderson, G. G., Zhang, Y., Lench, N. J., Carey, A., Cardon, L. R., Moffatt, M. F., and Cookson, W. O. (2001), 'Extent and distribution of linkage disequilibrium in three genomic regions', *Am. J. Hum. Genet.* **68**(1), 191–197.

Andolfatto, P. and Nordborg, M. (1998), 'The effect of gene conversion on intralocus associations', *Genetics* **148**(3), 1397–1399.

Austerlitz, F. and Heyer, E. (1998), 'Social transmission of reproductive behavior increases frequency of inherited disorders in a young-expanding population', *Proc. Natl. Acad. Sci. USA* **95**, 15140–15144.

Bahlo, M. and Griffiths, R. C. (2000), 'Inference from gene trees in a subdivided population', *Theor. Popul. Biol.* **57**(2), 79–95.

Bahlo, M. and Griffiths, R. C. (2001), 'Coalescence time for two genes from a subdivided population', *J. Math. Biol.* **43**(5), 397–410.

Baird, S. J. E., Barton, N. H., and Etheridge, A. M. (2003), 'The distribution of surviving blocks of an ancestral genome', *Theor. Popul. Biol.* **64**(4), 451–471.

Balding, D. J., Bishop, M. and Cannings, C., eds. (2001), *Handbook of Statistical Genetics*, John Wiley and Sons.

Barton, N. H., Depaulis, F., and Etheridge, A. M. (2002), 'Neutral evolution in spatially continuous populations', *Theor. Popul. Biol.* **61**(1), 31–48.

Beerli, P. and Felsenstein, J. (1999), 'Maximum-likelihood estimation of migration rates and effective population numbers in two populations using a coalescent approach', *Genetics* **152**(2), 763–773.

Beerli, P. and Felsenstein, J. (2001), 'Maximum likelihood estimation of a migration matrix and effective population sizes in n subpopulations by using a coalescent approach', *Proc. Natl. Acad. Sci. USA* **98**(8), 4563–4568.

Braverman, J. M., Hudson, R. R., Kaplan, N. L., Langley, C. H. and Stephan, W. (1995), 'The hitchhiking effect on the site frequency spectrum of DNA polymorphisms', *Genetics* **140**(2), 783–796.

Cann, R. L., Stoneking, M., and Wilson, A. C. (1987), 'Mitochondrial DNA and human evolution', *Nature* **325**(6099), 31–36.

Cardon, L. R. and Abecasis, G. R. (2003), 'Using haplotype blocks to map human complex trait loci', *Trends Genet.* **19**(3), 135–140.

Cardon, L. R. and Bell, J. I. (2001), 'Association study designs for complex diseases', *Nat. Rev. Genet.* **2**(2), 91–99.

Cavalli-Sforza, L.-L. (2001), *Genes, People and Language*, Penguin.

Cavalli-Sforza, L.-L., Menozzi, P., and Piazza, A. (1994), *The History and Geography of Human Genes*, Princeton University Press.

Chang, J. (1999), 'Recent common ancestor of all present human individuals', *Adv. Appl. Prob.* **31**, 1002–1026.

Charlesworth, B., Morgan, M. T., and Charlesworth, D. (1993), 'The effect of deleterious mutations on neutral molecular variation', *Genetics* **134**, 1289–1303.

Charlesworth, D. (2003), 'Effects of inbreeding on the genetic diversity of populations', *Philos Trans. R. Soc. Lond. B Biol. Sci.* **358** (1434), 1051–1070.

Chen, F. C. and Li, W. H. (2001), 'Genomic divergences between humans and other hominoids and the effective population size of the common ancestor of humans and chimpanzees', *Am. J. Hum. Genet.* **68**(2), 444–456.

Christiansen, F. B. (1999), *Population Genetics of Multiple Loci*, John Wiley and Sons.

Clark, A. G. (1997), 'Neutral behavior of shared polymorphism', *Proc. Natl. Acad. Sci. USA* **94**(15), 7730–7734.

Clayton, D. (2001), Population association, *in* D. J. Balding, M. Bishop, and C. Cannings, eds., 'Handbook of statistical genetics', John Wiley and Sons, pp. 519–540.

Daly, M. J., Rioux, J. D., Schaffner, S. F., Hudson, T. J., and Lander, E. S. (2001), 'High-resolution haplotype structure in the human genome', *Nat. Genet.* **29**(2), 229–232.

Derrida, B., Manrubia, S. C., and Zanette, D. H. (2000), 'On the genealogy of a population of biparental individuals', *J. Theor. Biol.* **203**(3), 303–315.

Devlin, B. and Roeder, K. (1999), 'Genomic control for association studies', *Biometrics* **55**(4), 997–1004.

Donnelly, K. (1983), 'The probability that related individuals share some section genome identical by descent', *Theor. Popul. Biol.* **23**, 34–63.

Donnelly, P. and Marjoram, P. (1989), 'The effect on genetic sampling distributions of correlations in reproduction', *Theor. Popul. Biol.* **35**, 22-35.

Donnelly, P., Nordborg, M., and Joyce, P. (2001), 'Likelihoods and simulation methods for a class of nonneutral population genetics models', *Genetics* **159**(2), 853–867.

Donnelly, P. and Tavaré, S. (1995), 'Coalescents and genealogical structure under neutrality', *Annu. Rev. Genet.* **29**, 401–421.

Donnelly, P. and Tavaré, S. (1997), *Progress in Population Genetics and Human Evolution*, Springer Verlag.

Donnelly, P., Tavaré, S., Balding, D. J., and Griffiths, R. C. (1996), 'Estimating the age of the common ancestor of men from the ZFY intron', *Science* **272**(5266), 1357–1359.

Drake, J. W., Charlesworth, B., Charlesworth, D. and Crow, J. F. (1998), 'Rates of spontaneous mutation', *Genetics* **148**(4), 1667–1686.

Durrett, R. (2002), *Probability Models for DNA Sequence Evolution*, Springer Verlag.

Ethier, S. N. and Griffiths, R. C. (1987), 'The infinitely-many-sites model as a measure-valued diffusion', *Ann. Probab.* **15**(2), 515–545.

Ethier, S. N. and Griffiths, R. C. (1990), 'The neutral two-locus model as a measure valued diffusion', *Adv. Appl. Prob.* **22**, 773–786.

Ewens, W. J. (1972), 'The sampling theory of selectively neutral alleles', *Theor. Popul. Biol.* **3**, 87–112.

Ewens, W. J. (2004), *Mathematical Population Genetics,* 2nd edn, Springer Verlag.

Eyre-Walker, A., Gaut, R. L., Hilton, H., Feldman, D. L., and Gaut, B. S. (1998), 'Investigation of the bottleneck leading to the domestication of maize', *Proc. Natl. Acad. Sci. USA* **95**(8), 4441–4446.

Fearnhead, P. (2003), 'Ancestral processes for non-neutral models of complex diseases', *Theor. Popul. Biol.* **63**(2), 115–130.

Fearnhead, P. and Donnelly, P. (2001), 'Estimating recombination rates from population genetic data', *Genetics* **159**(3), 1299–1318.

Fearnhead, P. and Donnelly, P. (2002), 'Approximate likelihood methods for estimating local recombination rates', *J. Roy. Stat. Soc. B* **64**, 657–680.

Felsenstein, J. (1981), 'Evolutionary trees from DNA sequences: a maximum likelihood approach', *J. Mol. Evol.* **17**(6), 368–376.

Felsenstein, J. (2003), *Inferring Phylogenies*, Sinauer.

Fisher, R. A. (1930), *The Genetical Theory of Natural Selection*, 1st edn, Clarendon Press.

Frisse, L., Hudson, R. R., Bartoszewicz, A., Wall, J. D., Donfack, J., and Di Rienzo, A. (2001), 'Gene conversion and different population histories may explain the contrast between polymorphism and linkage disequilibrium levels', *Am. J. Hum. Genet.* **69**(4), 831–843.

Fu, Y. X. (1994), 'Estimating effective population size or mutation rate using the frequencies of mutations of various classes in a sample of DNA sequences', *Genetics* **138**(4), 1375–1386.

Fu, Y. X. (1996), 'Estimating the age of the common ancestor of a DNA sample using the number of segregating sites', *Genetics* **144**(2), 829–838.

Fu, Y. X. (1997), 'Coalescent theory for a partially selfing population', *Genetics* **146**(4), 1489–1499.

Fullerton, S. M., Clark, A. G., Weiss, K. M., Nickerson, D. A., Taylor, S. L., Stengard, J. H., Salomaa, V., Vartiainen, E., Perola, M., Boerwinkle, E., and Sing, C. F. (2000), 'Apolipoprotein E variation at the sequence haplotype level: implications for the origin and maintenance of a major human polymorphism', *Am. J. Hum. Genet.* **67**(4), 881–900.

Gabriel, S. B., Schaffner, S. F., Nguyen, H., Moore, J. M., Roy, J., Blumenstiel, B., Higgins, J., DeFelice, M., Lochner, A., Faggart, M., Liu-Cordero, S. N., Rotimi, C., Adeyemo, A., Cooper, R., Ward, R., Lander, E. S., Daly, M. J., and Altshuler, D. (2002), 'The structure of haplotype blocks in the human genome', *Science* **296**(5576), 2225–2229.

Galtier, N., Depaulis, F., and Barton, N. H. (2000), 'Detecting bottlenecks and selective sweeps from DNA sequence polymorphism', *Genetics* **155**(2), 981–987.

Gibbs, R. A., Belmont, J. W., Hardenbol, P., Willis, T. D., Yu, F., Yang, H., Chang, L.-Y., Huang, W., Liu, B., and Snen, Y. (2003), 'The international hapmap project', *Nature* **426**, 789–796.

Glazko, G. V. and Nei, M. (2003), 'Estimation of divergence times for major lineages of primate species', *Mol. Biol. Evol.* **20**, 424–34.

Griffiths, R. C. (1980), 'Lines of descent in the diffusion approximation of neutral Wright–Fisher models', *Theor. Popul. Biol.* **17**(1), 37–50.

Griffiths, R. C. (1981), 'Neutral two-locus multiple allele models with recombination', *Theor. Popul. Biol.* **19**, 169–186.

Griffiths, R. C. (1987), 'Counting genealogical trees', *J. Math. Biol.* **25**(4), 423–431.

Griffiths, R. C. (1989), 'Genealogical-tree probabilities in the infinitely-many-site model', *J. Math. Biol.* **27**(6), 667–680.

Griffiths, R. C. (1991), 'Which locus has the oldest allele', *J. Math. Biol.* **29**(8), 763–777.

Griffiths, R. C. (1999), 'The time to the ancestor along sequences with recombination', *Theor. Popul. Biol.* **55**(2), 137–144.

Griffiths, R. C. and Marjoram, P. (1996), 'Ancestral inference from samples of DNA sequences with recombination', *J. Comput. Biol.* **3**(4), 479–502.

Griffiths, R. C. and Marjoram, P. (1997), An ancestral recombination graph, *in* P. Donnelly and S. Tavaré, eds., 'Progress in Population Genetics and Human Evolution', Springer Verlag, pp. 257–270.

Griffiths, R. C. and Tavaré, S. (1994), 'Simulating probability distributions in the coalescent', *Theor. Popul. Biol.* **46**, 131–159.

Griffiths, R. C. and Tavaré, S. (1995), 'Unrooted genealogical tree probabilities in the infinitely-many-sites model', *Math. Biosci.* **127**(1), 77–98.

Griffiths, R. C. and Tavaré, S. (1998), 'The age of a mutation in a general coalescent tree', *Stochastic Models* **14**, 273–295.

Gusfield, D. (1991), 'Efficient algorithms for inferring evolutionary trees', *Networks* **21**(1), 19–28.

Gusfield, D. (2001), 'Inference of haplotypes from samples of diploid populations: complexity and algorithms', *J. Comput. Biol.* **8**(3), 305–323.

Haldane, J. B. S. (1932), *The Causes of Evolution*, Longmans, Green & Co.

Hammer, M. F., Karafet, T. M., Redd, A. J., Jarjanazi, H., Santachiara-Benerecetti, S., Soodyall, H., and Zegura, S. L. (2001), 'Hierarchical patterns of global human Y-chromosome diversity', *Mol. Biol. Evol.* **18**(7), 1189–1203.

Hammer, M. F., Spurdle, A. B., karafet, T., Bonner, M. R., Wood, E. T., Novelletto, A., Malaspina, P., Mitchell, R. J., Horai, S., Jenkins, T., and Zegura, S. L. (1997), 'The geographic distribution of human y chromosome variation', *Genetics* **145**, 787–805.

Harding, R. M., Fullerton, S. M., Griffiths, R. C., and Clegg, J. B. (1997), 'A gene tree for beta-globin sequences from Melanesia', *J. Mol. Evol.* **44 Suppl 1**, 133–138.

Harris, E. E. and Hey, J. (1999), 'X chromosome evidence for ancient human histories', *Proc. Natl. Acad. Sci. USA* **96**(6), 3320–3324.

Hasegawa, M., Kishino, H., and Yano, T. (1985), 'Dating of the human-ape splitting by a molecular clock of mitochondrial DNA', *J. Mol. Evol.* **22**(2), 160–174.

Haubold, B., Kroymann, J., Ratzka, A., Mitchell-Olds, T., and Wiehe, T. (2002), 'Recombination and gene conversion in a 170-kb genomic region of Arabidopsis thaliana', *Genetics* **161**(3), 1269–1278.

Hein, J. (1990), 'Reconstructing evolution of sequences subject to recombination using parsimony', *Math. Biosci.* **98**(2), 185–200.

Hein, J. J. (1993), 'A heuristic method to reconstruct the history of sequences subject to recombination', *J. Mol. Evol.* **20**, 402–441.

Helgason, A., Hrafnkelsson, B., Gulcher, J. R., Ward, R., and Stefansson, K. (2003), 'A populationwide coalescent analysis of Icelandic matrilineal and patrilineal genealogies: evidence for a faster evolutionary rate of mtDNA lineages than Y chromosomes', *Am. J. Hum. Genet.* **72**(6), 1370–1388.

Hellmann, I., Ebersberger, I., Ptak, S. E., Paabo, S., and Przeworski, M. (2003), 'A neutral explanation for the correlation of diversity with recombination rates in humans', *Am. J. Hum. Genet.* **72**(6), 1527–1535.

Hey, J. and Wakeley, J. (1997), 'A coalescent estimator of the population recombination rate', *Genetics* **145**(3), 833–846.

Hill, W. G. and Robertson, A. (1966), 'The effect of linkage on limits to artificial selection', *Genet. Res.* **8**, 269–294.

Hudson, R. R. (1983*a*), 'Properties of a neutral allele model with intragenic recombination', *Theor. Popul. Biol.* **23**(2), 183–201.

Hudson, R. R. (1983*b*), 'Testing the constant-rate neutral allele model with protein sequence data', *Evolution* **37**(1), 213–217.

Hudson, R. R. (1987), 'Estimating the recombination parameter of a finite population model without selection', *Genet. Res.* **50**(3), 245–250.

Hudson, R. R. (1991), 'Gene genealogies and the coalescent process', *Oxford Surveys in Evolutionary Biology* **7**, 1–49.

Hudson, R. R. (2001*a*), Linkage disequilibrium and recombination, *in* D. J. Balding, M. Bishop, and C. Cannings, eds., 'Handbook of Statistical Genetics', John Wiley and Sons, pp. 309–324.

Hudson, R. R. (2001*b*), 'Two-locus sampling distributions and their application', *Genetics* **159**(4), 1805–1817.

Hudson, R. R. (2002), 'Generating samples under a Wright–Fisher neutral model of genetic variation', *Bioinformatics* **18**(2), 337–338.

Hudson, R. R. and Kaplan, N. L. (1985), 'Statistical properties of the number of recombination events in the history of a sample of DNA sequences', *Genetics* **111**(1), 147–164.

Hudson, R. R. and Kaplan, N. L. (1988), 'The coalescent process in models with selection and recombination', *Genetics* **120**(3), 831–840.

Hudson, R. R. and Kaplan, N. L. (1995), 'The coalescent process and background selection', *Philos. Trans. R. Soc. Lond. B Biol. Sci.* **349**(1327), 19–23.

Hudson, R. R., Kreitman, M., and Aguade, M. (1987), 'A test of neutral molecular evolution based on nucleotide data', *Genetics* **116**(1), 153–159.

Hudson, R. R., Slatkin, M., and Maddison, W. P. (1992), 'Estimation of levels of gene flow from DNA sequence data', *Genetics* **132**(2), 583–589.

Jeffreys, A. J., Holloway, J. K., Kauppi, L., May, C. A., Neumann, R., Slingsby, M. T., and Webb, A. J. (2004), 'Meiotic recombination hot spots and human DNA diversity', *Philos Trans. R. Soc. Lond. B Biol. Sci.* **359**(1441), 141–152.

Jeffreys, A. J., Kauppi, L. and Neumann, R. (2001), 'Intensely punctate meiotic recombination in the class II region of the major histocompatibility complex', *Nat. Genet.* **29**(2), 217–222.

Jobling, M. A., Hurles, M. E. and Tyler-Smith, C. (2004), *Human Evolutionary Genetics*, Garland Science.

Jobling, M. A. and Tyler-Smith, C. (2003), 'The human Y chromosome: an evolutionary marker comes of age', *Nat. Rev. Genet.* **4**(8), 598–612.

Jukes, T. H. and Cantor, C. R. (1969), Evolution of protein molecules, *in* H. N. Munroe, ed., 'Mammalian Protein Metabolism', Academic Press, pp. 21–132.

Kammerle, K. (1989), 'Looking forward and backwards in a bisexual Moran model', *J. Appl. Prob.* **27**, 880–85.

Kammerle, K. (1991), 'The extinction probability of descendants in bisexual models of fixed population size', *J. Appl. Prob.* **28**, 489–502.

Kaplan, N. and Hudson, R. R. (1985), 'The use of sample genealogies for studying a selectively neutral *m*-loci model with recombination', *Theor. Popul. Biol.* **28**(3), 382–396.

Kaplan, N., Hudson, R. R., and Iizuka, M. (1991), 'The coalescent process in models with selection, recombination and geographic subdivision', *Genet. Res.* **57**(1), 83–91.

Kaplan, N. and Morris, R. (2001), 'Issues concerning association studies for fine mapping a susceptibility gene for a complex disease', *Genet. Epidemiol.* **20**(4), 432–457.

Kauppi, L., Jeffreys, A. J. and Keeney, S. (2004), 'Where the crossovers are: recombination distributions in mammals', *Nat. Rev. Genet.* **5**(6), 413–424.

Kimura, M. (1968), 'Evolutionary rate at the molecular level', *Nature* **217**(129), 624–626.

Kimura, M. (1969), 'The number of heterozygous nucleotide sites maintained in a finite population due to steady flux of mutations', *Genetics* **61**(4), 893–903.

Kimura, M. (1980), 'A simple method for estimating evolutionary rates of base substitutions through comparative studies of nucleotide sequences', *J. Mol. Evol.* **16**(2), 111–120.

Kimura, M. (1983), *The neutral theory of molecular evolution*, Oxford University Press.

Kimura, M. and Crow, J. (1964), 'The number of alleles that can be maintained in a finite population', *Genetics* **49**, 725–738.

Kingman, J. F. C. (1982*a*), 'The coalescent', *Stoch. Process. Appl.* **13**, 235–248.

Kingman, J. F. C. (1982*b*), 'On the genealogy of large populations', *J. Appl. Prob.* **19A**, 27–43.

Kingman, J. F. C. (2000), 'Origins of the coalescent. 1974–1982', *Genetics* **156**(4), 1461–1463.

Kreitman, M. (1983), 'Nucleotide polymorphism at the alcohol dehydrogenase locus of *Drosophila melanogaster*', *Nature* **304**, 412–417.

Krings, M., Stone, A., Schmitz, R. W., Krainitzki, H., Stoneking, M., and Paabo, S. (1997), 'Neandertal DNA sequences and the origin of modern humans', *Cell* **90**(1), 19–30.

Krone, S. M. and Neuhauser, C. (1997), 'Ancestral processes with selection', *Theor. Popul. Biol.* **51**, 210–237.

Kuhner, M. K., Yamato, J., and Felsenstein, J. (1995), 'Estimating effective population size and mutation rate from sequence data using Metropolis-Hastings sampling', *Genetics* **140**(4), 1421–1430.

Kuhner, M. K., Yamato, J., and Felsenstein, J. (1998), 'Maximum likelihood estimation of population growth rates based on the coalescent', *Genetics* **149**(1), 429–434.

Kuhner, M. K., Yamato, J., and Felsenstein, J. (2000), 'Maximum likelihood estimation of recombination rates from population data', *Genetics* **156**(3), 1393–1401.

Lander, E. S., Linton, L. M., Birren, B., Nusbaum, C., Zody, M. C., Baldwin, J., Devon, K., Dewar, K., Doyle, M., FitzHugh, W., Funke, R., Gage, D., *et al.* (2001), 'Initial sequencing and analysis of the human genome', *Nature* **409**(6822), 860–921.

Larribe, F., Lessard, S., and Schork, N. J. (2002), 'Gene mapping via the ancestral recombination graph', *Theor. Popul. Biol.* **62**(2), 215–229.

Lessard, S. and Wakeley, J. (2004), 'The two-locus ancestral graph in a subdivided population: convergence as the number of demes grows in the island model', *J. Math. Biol.* **48**(3), 275–292.

Lewin, B. (2003), *Genes VIII*, Oxford University Press.

Li, N. and Stephens, M. (2003), 'Modeling linkage disequilibrium and identifying recombination hotspots using single-nucleotide polymorphism data', *Genetics* **165**(4), 2213–2233.

Liu, J. S., Sabatti, C., Teng, J., Keats, B. J., and Risch, N. (2001), 'Bayesian analysis of haplotypes for linkage disequilibrium mapping', *Genome Res.* **11**(10), 1716–1724.

Malécot, G. (1971), 'Structure géographique et variabilité d'une grande population', *Excerpta Medica* **III**(8), 138–154.

Marais, G. (2003), 'Biased gene conversion: implications for genome and sex evolution', *Trends Genet.* **19**(6), 330–338.

Martin, E. R., Lai, E. H., Gilbert, J. R., Rogala, A. R., Afshari, A. J., Riley, J., Finch, K. L., Stevens, J. F., Livak, K. J., Slotterbeck, B. D., Slifer, S. H., Warren, L. L., Conneally, P. M., Schmechel, D. E., Purvis, I., Pericak-Vance, M. A., Roses, A. D., and Vance, J. M. (2000), 'SNPing away at complex diseases: analysis of single-nucleotide polymorphisms around APOE in Alzheimer disease', *Am. J. Hum. Genet.* **67**(2), 383–394.

McVean, G. A. T. (2002), 'A genealogical interpretation of linkage disequilibrium', *Genetics* **162**(2), 987–991.

McVean, G. A. T., Myers, S. R., Hunt, S., Deloukas, P., Bentley, D. R., and Donnelly, P. (2004), 'The fine-scale structure of recombination rate variation in the human genome', *Science* **304**(5670), 581–584.

McVean, G., Awadalla, P., and Fearnhead, P. (2002), 'A coalescent-based method for detecting and estimating recombination from gene sequences', *Genetics* **160**(3), 1231–1241.

Mohle, M. (2000), 'Ancestral processes in population genetics-the coalescent', *J. Theor. Biol.* **204**(4), 629–638.

Molitor, J., Marjoram, P., and Thomas, D. (2003), 'Fine-scale mapping of disease genes with multiple mutations via spatial clustering techniques', *Am. J. Hum. Genet.* **73**(6), 1368–1384.

Moran, P. A. (1958*a*), 'A general theory of the distribution of gene frequencies. I. Non-overlapping generations', *Proc. R. Soc. Lond. B Biol. Sci.* **149**(934), 113–116.

Moran, P. A. (1958*b*), 'A general theory of the distribution of gene frequencies. II. Non-overlapping generations', *Proc. R. Soc. Lond. B Biol. Sci.* **149**(934), 113–116.

Moran, P. A. (1958*c*), 'The rate of approach to homozygosity', *Ann. Hum. Genet.* **23**(1), 1–5.

Morris, A. P., Whittaker, J. C., and Balding, D. J. (2004), 'Little loss of information due to unknown phase for fine-scale linkage-disequilibrium mapping with single-nucleotide-polymorphism genotype data', *Am. J. Hum. Genet.* **74**(5), 945–953.

Morris, A. P., Whittaker, J. C., and Balding, D. J. (2002), 'Fine-scale mapping of disease loci via shattered coalescent modeling of genealogies', *Am. J. Hum. Genet.* **70**(3), 686–707.

Morris, A. P., Whittaker, J. C., Xu, C.-F., Hosking, L. K., and Balding, D. J. (2003), 'Multi-point linkage-disequilibrium mapping narrows location interval and identifies mutation heterogeneity', *Proc. Natl. Acad. Sci. USA* **100**(23), 13442–13446.

Myers, S. R. and Griffiths, R. C. (2003), 'Bounds on the minimum number of recombination events in a sample history', *Genetics* **163**(1), 375–394.

Nagylaki, T. (1983), 'Evolution of a large population under gene conversion', *Proc. Natl. Acad. Sci. USA* **80**(19), 5941–5945.

Nagylaki, T. (1985), 'Biased intrachromosomal gene conversion in a chromosome lineage', *J. Math. Biol.* **21**(3), 215–235.

Nagylaki, T. (2002), 'When and where was the most recent common ancestor', *J. Math. Biol.* **44**(3), 253–275.

Nei, M. and Gojobori, T. (1986), 'Simple methods for estimating the numbers of synonymous and nonsynonymous nucleotide substitutions', *Mol. Biol. Evol.* **3**(5), 418–426.

Neuhauser, C. (1999), 'The ancestral graph and gene genealogy under frequency-dependent selection', *Theor. Popul. Biol.* **56**(2), 203–214.

Neuhauser, C. and Krone, S. M. (1997), 'The genealogy of samples in models with selection', *Genetics* **145**(2), 519–534.

Nielsen, R. (1998), 'Maximum likelihood estimation of population divergence times and population phylogenies under the infinite sites model', *Theor. Popul. Biol.* **53**(2), 143–151.

Nielsen, R. (2000), 'Estimation of population parameters and recombination rates from single nucleotide polymorphisms', *Genetics* **154**(2), 931–942.

Nielsen, R. and Signorovitch, J. (2003), 'Correcting for ascertainment biases when analyzing SNP data: applications to the estimation of linkage disequilibrium', *Theor. Popul. Biol.* **63**(3), 245–255.

Nielsen, R. and Wakeley, J. (2001), 'Distinguishing migration from isolation: a Markov chain Monte Carlo approach', *Genetics* **158**(2), 885–896.

Nordborg, M. (1997), 'Structured coalescent processes on different time scales', *Genetics* **146**(4), 1501–1514.

Nordborg, M. (1998), 'On the probability of Neanderthal ancestry', *Am. J. Hum. Genet.* **63**(4), 1237–1240.

Nordborg, M. (2000), Coalescent theory, *in* D. J. Balding, M. Bishop, and C. Cannings, eds., 'Handbook in Statistical Genetics', John Wiley and Sons, pp. 179–212.

Nordborg, M. and Donnelly, P. (1997), 'The coalescent process with selfing', *Genetics* **146**(3), 1185–1195.

Nordborg, M. and Tavaré, S. (2002), 'Linkage disequilibrium: what history has to tell us', *Trends Genet.* **18**(2), 83–90.

Notohara, M. (1990), 'The coalescent and the genealogical process in geographically structured population', *J. Math. Biol.* **29**(1), 59–75.

Oota, H., Settheetham-Ishida, W., Tiwawech, D., Ishida, T., and Stoneking, M. (2001), 'Human mtdna and y-chromosome variation is correlated with matrilocal versus patrilocal residence', *Nat Genet* **29**, 20–21.

Pannell, J. R. (2003), 'Coalescence in a metapopulation with recurrent local extinction and recolonization', *Evolution* **57**(5), 949–961.

Patil, N., Berno, A. J., Hinds, D. A., Barrett, W. A., Doshi, J. M., Hacker, C. R., Kautzer, C. R., Lee, D. H., Marjoribanks, C., McDonough, D. P., Nguyen, B. T., Norris, M. C., Sheehan, J. B., Shen, N., Stern, D., Stokowski, R. P., Thomas, D. J., Trulson, M. O., Vyas, K. R., Frazer, K. A., Fodor, S. P., and Cox, D. R. (2001), 'Blocks of limited haplotype diversity revealed by high-resolution scanning of human chromosome 21', *Science* **294**(5547), 1719–1723.

Pitman, J. (1999), 'Coalescents with multiple collisions', *Ann. Prob.* **27**, 1870–1902.

Pluzhnikov, A. and Donnelly, P. (1996), 'Optimal sequencing strategies for surveying molecular genetic diversity', *Genetics* **144**(3), 1247–1262.

Posada, D. and Wiuf, C. (2003), 'Simulating haplotype blocks in the human genome', *Bioinformatics* **19**(2), 289–290.

Pritchard, J. K. (2001), 'Are rare variants responsible for susceptibility to complex diseases', *Am. J. Hum. Genet.* **69**(1), 124–137.

Pritchard, J. K. and Cox, N. J. (2002), 'The allelic architecture of human disease genes: common disease-common variant... or not', *Hum. Mol. Genet.* **11**(20), 2417–2423.

Pritchard, J. K. and Przeworski, M. (2001), 'Linkage disequilibrium in humans: models and data', *Am. J. Hum. Genet.* **69**(1), 1–14.

Pritchard, J. K., Stephens, M., and Donnelly, P. (2000), 'Inference of population structure using multilocus genotype data', *Genetics* **155**(2), 945–959.

Pritchard, J. K., Stephens, M., Rosenberg, N. A., and Donnelly, P. (2000), 'Association mapping in structured populations', *Am. J. Hum. Genet.* **67**(1), 170–181.

Przeworski, M., Charlesworth, B., and Wall, J. D. (1999), 'Genealogies and weak purifying selection', *Mol. Biol. Evol.* **16**(2), 246–252.

Przeworski, M. and Wall, J. D. (2001), 'Why is there so little intragenic linkage disequilibrium in humans', *Genet. Res.* **77**(2), 143–151.

Ptak, S. E., Roeder, A. D., Stephens, M., Gilad, Y., Paabo, S., and Przeworski, M. (2004), 'Absence of the TAP2 Human Recombination Hotspot in Chimpanzees', *PLoS Biol.* **2**(6), E155.

Ptak, S. E., Voelpel, K., and Przeworski, M. (2004), 'Insights into recombination from patterns of linkage disequilibrium in humans', *Genetics* **167**(1), 387–397.

Rafnar, T., Thorlacius, S., Steingrimsson, E., Schierup, M. H., Madsen, J. N., Calian, V., Eldon, B. J., Jonsson, T., Hein, J., and Thorgeirsson, S. S. (2004), 'The Icelandic Cancer Project–a population-wide approach to studying cancer', *Nat. Rev. Cancer* **4**(6), 488–492.

Rannala, B. and Reeve, J. P. (2001), 'High-resolution multipoint linkage-disequilibrium mapping in the context of a human genome sequence', *Am. J. Hum. Genet.* **69**(1), 159–178.

Reich, D. E., Cargill, M., Bolk, S., Ireland, J., Sabeti, P. C., Richter, D. J., Lavery, T., Kouyoumjian, R., Farhadian, S. F., Ward, R., and Lander, E. S. (2001), 'Linkage disequilibrium in the human genome', *Nature* **411**(6834), 199–204.

Reich, D. E. and Lander, E. S. (2001), 'On the allelic spectrum of human disease', *Trends Genet.* **17**(9), 502–510.

Reich, D. E., Schaffner, S. F., Daly, M. J., McVean, G., Mullikin, J. C., Higgins, J. M., Richter, D. J., Lander, E. S., and Altshuler, D. (2002), 'Human genome sequence variation and the influence of gene history, mutation and recombination', *Nat. Genet.* **32**(1), 135–142.

Rosenberg, N. A. (2002), 'The probability of topological concordance of gene trees and species trees', *Theor. Popul. Biol.* **61**(2), 225–247.

Rosenberg, N. A. (2003), 'The shapes of neutral gene genealogies in two species: probabilities of monophyly, paraphyly, and polyphyly in a coalescent model', *Evolution* **57**(7), 1465–1477.

Rosenberg, N. A. and Nordborg, M. (2002), 'Genealogical trees, coalescent theory and the analysis of genetic polymorphisms', *Nat. Rev. Genet.* **3**(5), 380–390.

Rosenberg, N. A., Pritchard, J. K., Weber, J. L., Cann, H. M., Kidd, K. K., Zhivotovsky, L. A., and Feldman, M. W. (2002), 'Genetic structure of human populations', *Science* **298**(5602), 2381–2385.

Sabeti, P. C., Reich, D. E., Higgins, J. M., Levine, H. Z. P., Richter, D. J., Schaffner, S. F., Gabriel, S. B., Platko, J. V., Patterson, N. J., McDonald, G. J., Ackerman, H. C., Campbell, S. J., Altshuler, D., Cooper, R., Kwiatkowski, D., Ward, R., and Lander, E. S. (2002), 'Detecting recent positive selection in the human genome from haplotype structure', *Nature* **419**(6909), 832–837.

Satta, Y. and Takahata, N. (2002), 'Out of Africa with regional interbreeding? Modern human origins', *Bioessays* **24**(10), 871–875.

Saunders, I. W., Tavaré, S., and Watterson, G. (1984), 'On the genealogy of nested subsamples from a haploid population', *Adv. Appl. Probab.* **16**(3), 471–491.

Schierup, M. H. and Hein, J. (2000*a*), 'Consequences of recombination on traditional phylogenetic analysis', *Genetics* **156**(2), 879–891.

Schierup, M. H. and Hein, J. (2000*b*), 'Recombination and the molecular clock', *Mol. Biol. Evol.* **17**(10), 1578–1579.

Schierup, M. H., Mikkelsen, A. M., and Hein, J. (2001), 'Recombination, balancing selection and phylogenies in MHC and self-incompatibility genes', *Genetics* **159**(4), 1833–1844.

Schierup, M. H., Vekemans, X., and Charlesworth, D. (2000), 'The effect of subdivision on variation at multi-allelic loci under balancing selection', *Genet. Res.* **76**(1), 51–62.

Schierup, M. H., Vekemans, X., and Christiansen, F. B. (1998), 'Allelic genealogies in sporophytic self-incompatibility systems in plants', *Genetics* **150**(3), 1187–1198.

Schweinsberg, J. (2000), 'Coalescents with simultaneous multiple collisions', *Electron. J. Probab.* **5**, 1–50.

Seielstad, M. T., Minch, E., and Cavalli-Sforza, L. L. (1998), 'Genetic evidence for a higher female migration rate in humans', *Nat. Genet.* **20**, 278–280.

Semple, C. and Steel, M. (2003), *Phylogenetics*, Oxford University Press.

Simonsen, K. L. and Churchill, G. (1997), 'A Markov Chain model of coalescence with recombination', *Theor. Popul. Biol.* **52**, 43–59.

Slade, P. F. (2000), 'Most recent common ancestor probability distributions in gene genealogies under selection', *Theor. Popul. Biol.* **58**(4), 291–305.

Slade, P. F. (2001), 'Simulation of "hitch-hiking" genealogies', *J. Math. Biol.* **42**(1), 41–70.

Slatkin, M. (1991), 'Inbreeding coefficients and coalescence times', *Genet. Res.* **58**(2), 167–175.

Slatkin, M. (1996), 'Gene genealogies within mutant allelic classes', *Genetics* **143**(1), 579–587.

Slatkin, M. (2000), A coalescent view of population structure, *in* Singh and Krimbas, eds., 'Evolutionary Genetics', Cambridge University Press, pp. 418–429.

Slatkin, M. (2001), 'Simulating genealogies of selected alleles in a population of variable size', *Genet. Res.* **78**(1), 49–57.

Slatkin, M. and Hudson, R. R. (1991), 'Pairwise comparisons of mitochondrial DNA sequences in stable and exponentially growing populations', *Genetics* **129**(2), 555–562.

Slatkin, M. and Veuille, M. (2002), *Modern Developments in Theoretical Population Genetics*, Oxford University Press.

Sokal, R. R., Oden, N. L., and Thomson, B. A. (1999), 'A problem with synthetic maps', *Hum. Biol.* **71**(1), 1–13.

Song, Y. S. (2003), 'On the combinatorics of rooted binary phylogenetic trees', *Annals of Combinatorics* **7**, 365–379.

Song, Y. S. and Hein, J. (2003), Parsimonious reconstruction of sequence evolution and haplotype blocks: finding the minimum number of recombination events, *in* 'Lecture Notes in Bioinformatics, Proceedings of WABI'03', pp. 287–302.

Song, Y. S. and Hein, J. (2004*a*), 'On the minimum number of recombination events in the evolutionary history of DNA sequences', *J. Math. Biol.* **48**(2), 160–186.

Song, Y. S. and Hein, J. (2004*b*), 'Constructing minimal ancestral recombination graphs', *J. Comput. Biol.* **in press**.

Stefansson, H., Sarginson, J., Kong, A., Yates, P., Steinthorsdottir, V., Gudfinnsson, E., Gunnarsdottir, S., Walker, N., Petursson, H., Crombie, C., Ingason, A., Gulcher, J. R., Stefansson, K., and St Clair., D. (2003), 'Association of neuregulin 1 with schizophrenia confirmed in a Scottish population', *Am. J. Hum. Genet.* **72**(1), 83–87.

Stephens, M. (2000), 'Times on trees, and the age of an allele', *Theor. Popul. Biol.* **57**(2), 109–119.

Stephens, M. and Donnelly, P. (2000), 'Inference in molecular population genetics', *J. Roy. Stat. Soc. B* **62**(4), 605–635.

Stephens, M. and Donnelly, P. (2003), 'A comparison of Bayesian methods for haplotype reconstruction from population genotype data', *Am. J. Hum. Genet.* **73**(6), 1162–1169.

Stephens, M., Smith, N. J., and Donnelly, P. (2001), 'A new statistical method for haplotype reconstruction from population data', *Am. J. Hum. Genet.* **68**(4), 978–989.

Strachan, T. and Read, A. P. (2003), *Human Molecular Genetics 3*, BIOS Scientific Publishers Ltd, John Wiley and Sons.

Stumpf, M. P. and Goldstein, D. B. (2001), 'Genealogical and evolutionary inference with the human Y chromosome', *Science* **291**(5509), 1738–1742.

Stumpf, M. P. H. (2002), 'Haplotype diversity and the block structure of linkage disequilibrium', *Trends Genet.* **18**(5), 226–228.

Stumpf, M. P. H. and McVean, G. A. T. (2003), 'Estimating recombination rates from population-genetic data', *Nat. Rev. Genet.* **4**(12), 959–968.

Tajima, F. (1983), 'Evolutionary relationship of DNA sequences in finite populations', *Genetics* **105**(2), 437–460.

Tajima, F. (1989), 'Statistical method for testing the neutral mutation hypothesis by DNA polymorphism', *Genetics* **123**(3), 585–595.

Takahata, N. (1988), 'The coalescent in two partially isolated diffusion populations', *Genet. Res.* **52**(3), 213–222.

Takahata, N. (1990), 'A simple genealogical structure of strongly balanced allelic lines and trans-species evolution of polymorphism', *Proc. Natl. Acad. Sci. USA* **87**(7), 2419–2423.

Takahata, N., Lee, S. H., and Satta, Y. (2001), 'Testing multiregionality of modern human origins', *Mol. Biol. Evol.* **18**(2), 172–183.

Tavaré, S. (1984), 'Line-of-descent and genealogical processes, and their applications in population genetics models', *Theor. Popul. Biol.* **26**(2), 119–164.

Tavaré, S. (in press), *Ancestral Inference in Population Genetics*, Springer Verlag.

Tavaré, S., Balding, D. J., Griffiths, R. C., and Donnelly, P. (1997), 'Inferring coalescence times from DNA sequence data', *Genetics* **145**(2), 505–518.

Terwilliger, J. D., Haghighi, F., Hiekkalinna, T. S., and Goring, H. H. H. (2002), 'A bia-sed assessment of the use of SNPs in human complex traits', *Curr. Opin. Genet. Dev.* **12**(6), 726–734.

Uyenoyama, M. K. (1997), 'Genealogical structure among alleles regulating self-incompatibility in natural populations of flowering plants', *Genetics* **147**(3), 1389–1400.

Vekemans, X. and Slatkin, M. (1994), 'Gene and allelic genealogies at a gametophytic self-incompatibility locus', *Genetics* **137**(4), 1157–1165.

Venter, J. C., Adams, M. D., Myers, E. W., Li, P. W., Mural, R. J., Sutton, G. G., Smith, H. O., Yandell, M., Evans, C. A., Holt, R. A., *et al.* (2001), 'The sequence of the human genome', *Science* **291**(5507), 1304–1351.

Wakeley, J. (1996), 'Distinguishing migration from isolation using the variance of pairwise differences', *Theor. Popul. Biol.* **49**(3), 369–386.

Wakeley, J. (1997), 'Using the variance of pairwise differences to estimate the recombination rate', *Genet. Res.* **69**(1), 45–48.

Wakeley, J. (1998), 'Segregating sites in Wright's island model', *Theor. Popul. Biol.* **53**(2), 166–174.

Wakeley, J. (2000), 'The effects of subdivision on the genetic divergence of populations and species', *Evolution* **54**(4), 1092–1101.

Wakeley, J. and Aliacar, N. (2001), 'Gene genealogies in a metapopulation', *Genetics* **159**(2), 893–905.

Wakeley, J. and Takahashi, T. (2003), 'Gene genealogies when the sample size exceeds the effective size of the population', *Mol. Biol. Evol.* **20**(2), 208–213.

Wall, J. D. (2000), 'A comparison of estimators of the population recombination rate', *Mol. Biol. Evol.* **17**(1), 156–163.

Wall, J. D. (2001), 'Insights from linked single nucleotide polymorphisms: what we can learn from linkage disequilibrium', *Curr. Opin. Genet. Dev.* **11**(6), 647–651.

Wall, J. D. (2003), 'Estimating ancestral population sizes and divergence times', *Genetics* **163**(1), 395–404.

Wall, J. D. and Hudson, R. R. (2001), 'Coalescent simulations and statistical tests of neutrality', *Mol. Biol. Evol.* **18**(6), 1134–1135.

Wall, J. D. and Przeworski, M. (2000), 'When did the human population size start increasing?', *Genetics* **155**, 1865–1874.

Waterston, R. H., Lindblad-Toh, K., Birney, E., Rogers, J., Abril, J. F., Agarwal, P., Agarwala, R., and Ainscough, E. A. (2002), 'Initial sequencing and comparative analysis of the mouse genome', *Nature* **420**(6915), 520–562.

Watterson, E. A. (1975), 'On the number of segregating sites in genetical models without recombination', *Theor. Popul. Biol.* **7**(2), 256–276.

Weiss, K. M. and Terwilliger, J. D. (2000), 'How many diseases does it take to map a gene with SNPs', *Nat. Genet.* **26**(2), 151–157.

Westendorp, R. G. and Kirkwood, T. B. (1998), 'Human longevity at the cost of reproductive success', *Nature* **396**(6713), 743–746.

Wilkins, J. F. and Wakeley, J. (2002), 'The coalescent in a continuous, finite, linear population', *Genetics* **161**(2), 873–888.

Wilson, I. J., Weale, M. E., and Balding, D. J. (2003), 'Inferences from dna data: population histories, evolutionary processes and forensic match probabilities', *J. Roy. Stat. Soc. A Stat.* **166**(2), 155–188.

Wiuf, C. (2000*a*), 'A coalescence approach to gene conversion', *Theor. Popul. Biol.* **57**(4), 357–367.

Wiuf, C. (2000*b*), 'On the genealogy of a sample of neutral rare alleles', *Theor. Popul. Biol.* **58**(1), 61–75.

Wiuf, C. (2001), 'Rare alleles and selection', *Theor. Popul. Biol.* **59**(4), 287–296.

Wiuf, C. (2002), 'On the minimum number of topologies explaining a sample of DNA sequences', *Theor. Popul. Biol.* **62**(4), 357–363.

Wiuf, C. (2003), 'Inferring population history from genealogical trees', *J. Math. Biol.* **46**(3), 241–264.

Wiuf, C., Christensen, T., and Hein, J. (2001), 'A simulation study of the reliability of recombination detection methods', *Mol. Biol. Evol.* **18**(10), 1929–1939.

Wiuf, C. and Donnelly, P. (1999), 'Conditional genealogies and the age of a neutral mutant', *Theor. Popul. Biol.* **56**(2), 183–201.

Wiuf, C. and Hein, J. (1997), 'On the number of ancestors to a DNA sequence', *Genetics* **147**(3), 1459–1468.

Wiuf, C. and Hein, J. (1999*a*), 'A contribution to the discussion of J. Chang's paper 'Recent common ancestor of all present human individuals'', *Adv. Appl. Prob.* **31**, 1027–1031.

Wiuf, C. and Hein, J. (1999*b*), 'Recombination as a point process along sequences', *Theor. Popul. Biol.* **55**(3), 248–259.

Wiuf, C. and Hein, J. (1999*c*), 'The ancestry of a sample of sequences subject to recombination', *Genetics* **151**(3), 1217–1228.

Wiuf, C. and Hein, J. (2000), 'The coalescent with gene conversion', *Genetics* **155**(1), 451–462.

Wiuf, C. and Posada, D. (2003), 'A coalescent model of recombination hotspots', *Genetics* **164**(1), 407–417.

Wright, S. (1931), 'Evolution in Mendelian populations', *Genetics* **16**, 97–159.

Wright, S. (1951), 'The genetical structure of populations', *Ann. Eugenics* **15**, 323–354.

Xiong, M., Zhao, J., and Boerwinkle, E. (2003), 'Haplotype block linkage disequilibrium mapping', *Front. Biosci.* **8**, A85–93.

Yang, Z. (2002*a*), 'The estimation of ancestral population sizes in hominoids using data from multiple loci', *Genetics* **162**, 1811–1823.

Yang, Z. (2002*b*), 'Likelihood and Bayes estimation of ancestral population sizes in hominoids using data from multiple loci', *Genetics* **162**(4), 1811–1823.

Zhang, W., Collins, A., Maniatis, N., Tapper, W., and Morton, N. E. (2002), 'Properties of linkage disequilibrium (LD) maps', *Proc. Natl. Acad. Sci. USA* **99**(26), 17004–17007.

Zhao, Z. (2001), 'Chinese genealogies as a source for demographic research: A further assessment of their reliabilities and biases', *Population Studies* **55**, 181–193.

Index